CW00518896

IN HIS OWN WORDS

George Dowty, c.1927

In His Own Words

The Autobiography of
Sir George Dowty

THE HOBNOB PRESS

First published in the United Kingdom in 2020

by The Hobnob Press,
8 Lock Warehouse,
Severn Road, Gloucester GL1 2GA
www.hobnobpress.co.uk

© the Estate of Sir George Dowty, 2020

All rights reserved. No part of this publication may be reproduced, stored in a retrieval system, or transmitted in any form or by any means, electronic, mechanical, photocopying, recording or otherwise, without the prior permission of the publisher and copyright holder.

British Library Cataloguing in Publication Data
A catalogue record for this book is available from the British Library

ISBN 978-1-906978-94-5 (paperback)
      978-1-906978-95-2 (casebound)

Edited by Ally McConnell
Typeset in Scala 11/14 pt.
Typesetting and origination by John Chandler

CONTENTS

Preface *by George Dowty* 7

Foreword 9

Early Years 13

Starting in Business 32

The Business goes Public 44

The War Years 51

Post War 73

The Business Expands 95

Nationalised Industries 121

Education 126

Sport and Recreation 133

Family Life and General 142

Later Life 145

Index 153

PREFACE

by George Dowty

A FEW YEARS AGO I came across a poor photocopy of a typed transcript of what appeared to be my Father's autobiography, which he must have dictated to his secretary before his death in 1975. On closer inspection I realised that this was my Father's truly inspirational life story in his own words. Starting with nothing but a torrent of ideas it describes how he came to develop these into a world-class engineering enterprise.

Above all my Father was an inventor. He was never afraid to try to develop new ideas and designs. Many of these failed but his determination saw him through, and resulted in the development of products which were at the forefront of the aviation, mining and hydraulics businesses.

This book charts the rise of the son of a Pershore chemist who gained experience as a humble draughtsman before setting out on his own with almost no backing. These were exceptional times at the infancy of aviation which was to develop at a frantic pace through the 1920s, the Second World War and beyond.

I would like to express my sincere thanks to John Chandler and Ally McConnell at Gloucestershire Heritage Hub for kindly editing and encouraging the publishing of this book.

Whilst the world is now a very different place to my Father's era I would hope that young engineers and entrepreneurs out there can gain inspiration from this story. It was sheer hard work and determination that enabled my Father to success. The family motto, 'Labor Vincit' bears testimony to this.

Foreword

I STARTED my engineering business in 1930 with just £50 to my name. Many friends said I was crazy in throwing up a safe job as a draughtsman earning £5 a week. But had I possessed £5,000 then, the struggle that awaited me would have been the same. The years that followed were those of the depression and great unemployment. My lack of money forced me to improvise – to do without and to operate with the greatest economy.

Looking back at this distance, I feel it is no real hardship to be short of cash when starting an independent career – provided you have the stamina to withstand difficult and rigorous times.

I was a young man without any formal training in the arts and crafts. I was refused help by almost everyone to whom I turned. Virtually penniless – what had I to offer? I shall try to spell out, in this book, a life story of tenacity, perseverance and inventiveness.

This book is about the development of a large engineering enterprise from its grass roots. All the incidents, many of which took place nearly half a century ago, have been recalled with accuracy. I am fortunate in having kept comprehensive records and I have been able, where necessary, to refer to the two books written by my friend, the late Mr. L. T. C. Rolt: Parts 1 and 2 of *The Dowty Story*.

In the small Worcestershire town where I was born, I was brought up in a home which had to practice a real economy. This undoubtedly had its effect on my later thinking and behaviour.

The building up of a large industrial enterprise from nothing to one with annual sales of £120,000,000 demands a single-minded devotion to business. There is much satisfaction in the process – the thrill of the first order, the employment of your first worker, the purchase of the first machine tool. I found life a great adventure as I woke each morning.

Progress in business is often the result of invention and the impulse to make things better; but this must be followed by initiative and determination.

History has demonstrated that the most notable winners usually encountered heartbreaking obstacles before they triumphed. They finally won because they refused to be discouraged by their defeats. Inventors get ideas literally from out of the air; in my case ideas have come at unusual times and places. I have found that music can stimulate my thinking; sometimes in the theatre, quite unexpectedly, has come the flash of an idea.

In the early days of my business I went round every drawing board every day. There were countless dimensions on these drawings. Then, travelling on the London underground one day, I suddenly became conscious of a wrong dimension on one of those drawings. When I returned I discovered to my surprise that the error was indeed a fact. I seemed to have powers of observation for detecting errors in design that many told me were uncanny.

I'm exasperated by people who convince themselves something will not work and won't even try. What would my company have lost if I'd accepted the widely-held theory that liquids were incompressible?

I take the view that there are few products on the market that cannot be improved. You don't have to be a genius to see this. Sometimes this work can be as rewarding as thinking up a new product. The successful man must always seek better ways of doing things, attempting methods that are not in the book.

Opportunities don't necessarily come by spending money lavishly. Lord Rutherford, responsible for splitting the atom, said he didn't spend more than £100 on any one experiment. Tommy Sopwith, the aviation pioneer, built an aeroplane and its engine for £1,000, saying, 'I had to, that was all I had. '

Successful business is the essential ingredient which makes up the economy of an industrial nation. But it is an art and not a science. It is as much an art as composing music or producing a painting. Our educationalists would deny this but then, they only theorize and have little if any practical experience.

In taking up engineering I had no-one to whom I could turn for advice. The on-training I had was at the factories and offices in which I worked. I was entirely self-taught. But I had one priceless gift – a flair for invention.

For years I tried with increasing frustration to get my ideas taken up. But when somebody showed some interest it usually came to nothing. Tenaciously I kept plugging on. When I put forward proposals for setting up my own business I found no-one to back me. Never once did I think of going it alone; in the end fate decided otherwise.

Many people day-dream of starting their own businesses from scratch but few do it. This is understandable. To give up a steady job to become your own boss means substituting uncertainty for security. It inevitably involves greater responsibility and more worry. In my case, the setting up of a company in a time of depression was an exceedingly shaky matter.

Even that man Melville, whose London office I used in the very early days in return for an interest in my business, lost confidence when I had completed my first solitary order from Japan and pulled out with a few hundred pounds profit. At the time I badly needed that money in the business. But those I despised most were the bankers who refused me the modest help needed so desperately as I tentatively took my first steps.

There are, it seems to me, two extremes in business. One I liken to the building of the 'Queen Mary' – and the other to the purveying of fish and chips. In the former case, millions of pounds are required from governments or other financiers, and the work in progress builds up over many years. After the ship has been built, then where does the ship-builder look for the next order? Such a business can only be undertaken by those who must wait a long, long time for any return on their investment. On the other hand, the fish and chip merchant can sell his goods before he has paid for his raw material.

I've always said that the closer one can get to that type of business the better. That was how Sir Oliver Simmonds with his lock nut with the fibre insert – made a fortune.

The important people, I maintain, are the entrepreneurs. They are the innovators, those who have the ability to develop new products or new processes. Those who operate machines or hew coal, as well as the educationalists, lawyers, civil servants, shopkeepers and sportsmen, all owe their livelihoods to the entrepreneurs.

Yet, ironically and unreasonably, because of their numbers and the political considerations, it is possible for a section like the coal-miners to hold the country to ransom and claim that, as producers, they are more important than the creators of new businesses.

It has to be said, I was more interested in technical development than securing an adequate return for my work. It's a sad reflection that I made more money selling some land I had bought twenty years previously for raising bloodstock than I received for founding and building-up the Dowty Group and giving the company the use of my inventions.

Early Years

Pershore, a small Georgian town in Worcestershire with a splendid Norman abbey dating to 689 AD , is situated on the banks of the Stratford Avon that winds its way around Bredon Hill. Every third hour the ancient bells play tunes such as 'The Warrior Boy to the War Has Gone', which could be heard throughout the countryside. At the turn of the century, like many other country towns, lighting was by gas, its water pumped from wells and the only road transport was a horse drawn vehicle or bicycle. There was no public library.

It was here in the early morning of April 27th 1901, that twin sons were born to William and Laura Dowty. As a chemist and druggist my father's business was small indeed to support seven sons and a daughter. I was the seventh son by half an hour. Edward Flexton, my twin, and I were so alike in every particular that on our right feet we both had a hammer toe. It was a coincidence that we were given names with the initials following our surname in alphabetical order, DEF GH

It was a secluded life in which one had to make one's own amusements. I remember Chipperfield's Circus that visited the town each year with its silent cine films, and Stricklands Roundabouts came from Worcester every June for Pershore Fair. How I loved the smell of the steam engines. Numerous stalls lined the streets selling sweetmeats, toys, goldfish and so on.

In August there was a flower show and sports. I have a photograph of the 220 yards Handicap for Youths in 1907 with my twin brother, then aged 6, finishing a good second.

We both had bicycles and a favourite pastime was for one of us to be blindfold and to ride on the carrier over the back wheel. The peddler would circle round and round and then make off for a mile and so, and then ask the blindfold one where he was. We used to go to an attic window where we could see in the distance the Great Western trains travelling between Fladbury and Pershore stations. With the time table we would make records of the trains' time keeping. In summer time

The Dowty family in Pershore

there was swimming and fishing in the Avon, rambles over Bredon Hill, bicycling through the villages of Wyre, Cropthorne, and Elmley Castle.

When we could get others to join us there were paper chases and once a year, as we had no money for holidays, we went to my Uncle Tom Fisher's farm at Charingworth. He was married to my mother's sister Polly. I was 16 before I caught my first glimpse of the sea when I bicycled to North Devon to visit my brother, Joe, who was a chemist at Northam near Westward Ho!

Ours was a poor but happy home. My father was a Latin scholar and a lay preacher. I remember many Sundays walking with him across the fields to a neighbouring village where he conducted the service. His family came from Honeybourne, not far from Evesham. My grandfather, John Goddard Dowty, was a wheelwright and my great-great-great grandfather was John Dowty who died in 1799. A relation, the Reverend George Dowty, is buried in the Devonshire hamlet of Stockleigh English where he was vicar. Visiting there with my son we found evidence of the regard in which he was held, the stained glass window over the altar and the bells erected in his memory. In the vestry we found his prayer book which the vicar kindly gave us.

My mother was a Masters, a family long established in the neighbouring town of Evesham. Her father, brother and nephew had all been mayors of that town.

I do not remember a great deal about my father although there was an incident when I was walking with him one evening in 1910 along Station Road, Pershore, to see Hayley's Comet. After being told it would return in 1987 my father said that he would see it from the other side!

As a child I was always fascinated by the sky at night. I was told the farthest object seen in the universe was ten billion light years away and as light travels at over 186,000 miles a second my mind boggled at these awesome distances. I found it both incredible and frightening.

One evening in 1912 I set light to a bottle of magnesium powder used by my father for night photography. The glass bottle exploded in my face with the result that I lost my right eye. A rather stupid experiment!

At first the loss of an eye brought its problems and I found myself colliding on my right side with such things as lamp posts! I remember how painful these contacts were but gradually I became accustomed to this disability and it never interfered with my sports.

My father died from a stroke in 1913 and his estate amounted to but a few hundred pounds. My eldest brother, who was at that time working in Penang, came home to look after the family business, but my twin brother and I found life difficult. In those days when we had to receive punishment for any misdeeds we were always given gardening chores and this, I think, has resulted in my revulsion to land work all my life.

After my father's death l was befriended by my brother in law, Sidney Fell, a Worcester solicitor. An officer in the Worcestershire Regiment, he was killed in the First World War. He was only the second person I knew who owned an automobile. The other was my Uncle Clem Newey, a manufacturer of hooks and eyes and other fasteners, with factories in Birmingham. It was my brother in law who first interested me in engineering and he gave me a book called *The Wonders of the Engineer* and also a model steam engine. In our large family there had been no money for expensive toys and this was the first one of any consequence that I had owned.

Nowadays a child's creative talents can be inhibited by the wealth of ready-made toys but in my case necessity was the mother of invention. I can remember how I contrived a miniature set of fairground gallopers operated by the steam engine, making use of an old umbrella frame. Making toys stimulated imagination and initiative and no doubt helped me in my inventiveness. Those days were made most exciting by man's conquest of the air, Bleriot's flight across the Channel and the *Daily Mail* sponsored London to Manchester Air Race. My twin brother and I were always to be seen flying gliders or model aeroplanes from our bedroom window,

In 1913 I had my first model aeroplane given to me by my brother in law. This was a bi-plane with an elastic driven propeller and this was the first power-driven model I had ever had and it flew beautifully. I remember what a thrill it was to possess such a model. It was about this time too that I saw my first aeroplane, the visit of a Scandinavian airman called Gustav Hamel in a Bleriot monoplane that came to Pershore. Unfortunately, this pilot later lost his life crossing the Irish Sea.

My brother in law had a pianola and as my twin brother was a good pianist we undertook the major activity of making a pianola roll. My brother discovered the holes that connected to each piano note and with the aid of a razor blade I cut slots in a band made from strips of thin brown paper glued together. The piece of music was "Grand Choir in G Major" by Alexander Dumas [probably Alexander Guilmant], one time organist of Notre Dame in Paris. It was a complicated piece but the finished roll was a work of art and gave excellent results.

There was a way in which as twins we were different. My brother was a pianist and organist of some merit. He studied under Ivor Atkins (later Sir Ivor), organist of Worcester Cathedral. He later was organist at St. Mary's Worcester and frequently assisted other churches. Sir Edward Elgar was a Worcester man and as a boy I remember the music shop near the cathedral run by the Elgar family.

We had an excellent early education at a small private school in Pershore at Gore House, Bridge Street, run by the Brickell family with the elderly Mrs. Brickell taking mathematics and her daughters Edith and Nellie taking other subjects. It was due to the good tuition I had at this private school that I found myself a match for any of the scholars when I entered Worcester Royal Grammar School in 1913. Each day a three-mile bicycle ride to Pershore Station, a train to Worcester, and a two-mile walk to school. This journey day and night through all weathers would not be tolerated by young people today.

As identical twins my brother and I were always creating problems so that at school we soon had to be separated, being placed in different forms. Twice in the school's cross country race we were placed first and second but no-one could tell which had come in first.

I have my school reports which were good. In every subject were remarks such as 'excellent' or 'good' and in 1914 I won a form prize.

During that year the Perrins Hall was built through the generosity of Mr. Dyson Perrin of Worcester Sauce fame.

The headmaster was Mr. Hillard and for a year the acting head was 'Tubby' Carter, a great character who taught us chemistry. Carter was a great vocalist and I can remember his splendid rendering of 'Take a Pair of Sparkling Eyes'.

I achieved some notoriety by throwing a sixth-former's cap down the lift shaft on Foregate Street Station, Worcester. He must have annoyed me in some way but, of course, I paid the penalty by receiving six strokes of the best and losing a merit holiday. From my reports this incident occurred in the spring term of 1915.

My brother and I left school at the end of the winter term of 1915 to look after some of my brother's businesses for during the First World War conscription was introduced.

World War One started in August 1914 and I remember the day for my mother, my twin brother and I went on a picnic to a place called Gig Bridge, a short walk from the town. In those days we took the *Daily Mail* and I recall our concern at the murders in Serbia which resulted in the outbreak of war. After twelve months working in Pershore for the business of my brother Robert I was not at all happy and as my eldest brother John felt that he could look after both those businesses it gave me the opportunity to start work at Heenan and Froude Engineers in Worcester in January 1917. None of my family – even my twin brother – shared my interests, so when I started my engineering career I did so without friends or influence and entered my first factory as just another boy. Every morning and evening when on my way to school I had passed by this engineering factory outside of Worcester's Shrub Hill Station and was intrigued with what I saw going on through the windows. In these days that company was building the Smith Radial Aero Engine and the Constantinesco Interrupter Gear, a hydraulic device that permitted the firing of bullets through the rotating blades of an aeroplane propeller.

I started each morning at 6.00 a.m. and worked eight and a half hours a day, on Saturdays until 12.00 noon. It was difficult getting up so early especially on winter mornings. I was paid six shillings a week.

Having only one eye I was nicknamed Nelson. Inside the factory gates were some railway tracks which I had to cross to get to my place of work. Twice in the morning darkness I tripped over these lines breaking my precious Thermos flask and losing my breakfast tea.

I remember the bawdy limericks in the toilets, particularly one about the works manager, a man called Rackham. Some years later this man visited me when my business was considerable but I doubt if he remembered me as the boy in his factory so many years previously.

My first job was pressure testing the cast iron aero cylinders by means of a hydraulic hand pump. So from my earliest introduction to engineering I was associated with hydraulics. Another of my jobs was tapping threads in small blind holes in the engine cylinders. I recall the number of cylinders I scrapped by breaking taps in the blind holes. This experience I never forgot and in later years when responsible for design, I never permitted the tapping of blind holes.

I was only a boy in the factory with no-one to help or to give advice for there was no apprenticeship in those wartime days. I had an element of luck in the benevolent interest of a cheerful individual in blue overalls who always seemed to be hovering near the front door. He had sufficient influence to arrange for my transfer from one department to another.

I took evening classes in mechanical engineering at the Victoria Institute in the Foregate paying my own fees and I took a postal course on the Internal Combustion Engine. After a few months I was transferred to the inspection department and then to the drawing office where I learned to draw the hard way, but it taught me how to put my ideas on to paper. Even in those immature days I had ideas including one for aerial navigation not dissimilar to those in use today.

Being a twin had its problems for there were times when I was accused of being about the city when my clock card showed I was at work! In those days I lived with my mother in a house in Sansome Walk almost opposite St. Mary's Church. The vicar was the Reverend Frost who had two attractive daughters who my brother and I had our eyes on.

I was for some time a member of the Worcester Rowing Club. Its headquarters were on the Pitchcroft. I frequently took boats on the Severn but never realised how easy it was to get into trouble, neither had I been warned of the hazards to be avoided. One day I took my boat too near a weir with the result that it was sucked against the weir over which water was cascading. My shouts aroused the lock keeper who came to my assistance for the boat was rapidly filling with water. I well remember this very frightening experience.

Answering an advertisement in *Flight* in July 1918 I obtained a job in London as a draughtsman with the British Aerial Transport Company of Hythe Road, Willesden Junction at a salary of 35s. a week. From that day I became self supporting. I paid 30s. a week for my digs, leaving 5s. for clothes, railway fares home, etc. In my evenings I studied and designed a potato digger and an internal combustion turbine for I was even then concerned with the loss of efficiency in the starting and stopping of the pistons of an internal combustion engine twice in each revolution.

Two men helped me greatly, Mr. Wynn who came from Belliss and Morcom Ltd and later retired to New Zealand, and the late Mr. R. B. Elliott: I shall always remember the encouragement they gave to an inexperienced young man. It was here that I had my first introduction to aircraft design, particularly on undercarriages, which undoubtedly stimulated my interest in these products that were later to be my life's work.

The managing direcor, Fritz Koolhoven, a Dutchman, was of great stature. I remember seeing him chase his attractive wife round the factory brandishing a revolver.

Without money for entertainment I taught myself calculus and in May 1919 I took my first flight, a joy ride in an Avro 504K.

The chief designer, Bobby Noorduyn, came down to my home at Worcester one weekend to visit Malvern to buy a Morgan three wheeler. He remained my good friend until his death in Canada where he had been a successful aircraft designer. The aeroplanes designed at that time were the Bat 'Bantam' and Bat 'Basilisk', and the FK 27, the first civil aeroplane.

In October 1918 I visited the Enemy Aircraft Exhibition at the Agricultural Hall, Islington and later that month took my mother to see the play 'Chu Chin Chow' at His Majesty's Theatre. About that time I saw the plays 'The Lilac Domino', 'Going Up', and 'The Count of Luxembourg', all good musical comedies. I spent my Saturdays walking about London and visiting places I had heard so much about. I bought myself one meal at Lyons Corner House off Piccadilly. It was surprising one could buy so much in those days for eight pence.

The war at an end it seemed to me that the smaller aircraft companies could not survive. In January 1919 I applied to the aviation division of Boulton Paul, Norwich, for a position advertised but was unsuccessful. It was a twist of fate that some years later I was to buy that business.

I left the British Aerial Transport company in June 1919 to join Thomas Cook and Sons of York, to work on naval range finders. In some ways I enjoyed my stay there. The factory was closed on race days and I remember putting a small wager on a horse called "Race Rock" that won the Ebor Handicap. It is surprising how punters always remember their winners!

I found York a most interesting city and often walked round the walls which almost encircle the city. The Minster is a splendid building of great architectural beauty.

Not liking the work I left York in September. I had an unhappy experience on my leaving. A man with whom I shared lodgings borrowed £10 from me, which represented two weeks' salary. He was most plausible and said he would come to the station to see me away and repay the loan. He never turned up. I suppose these were lessons I had to learn.

I took a position with the British Portland Cement Company of Lloyds Avenue in the City of London. I was engaged on equipping a factory in India, work that did not interest me. I remember being sent to take an indicator diagram of a steam engine at their Irithlingborough Works. The only experience I had of such work was at the Victoria Institute in Worcester. I then joined the Dunlop Rubber Company at their Manor Mills factory in Birmingham in January 1920 and worked on the design of moulds for the manufacture of aircraft tyres and golf balls.

It was while I was at Dunlops that I heard of the death of the only grandparent I ever knew – my mother's mother. She was 98 and her death caused by a fall when from her bed she thought she saw a needle on the floor and fell in her attempt to retrieve it.

Ten months later I moved again to Rownson Drew & Clydesdale of Kings Cross, manufacturers of elevators and conveyors, and my design work was assembling products from standard catalogue items. Most uninteresting.

From the day I started my engineering career I was always critical of the way things were done and full of ideas for improvements. It was perhaps natural in those days that no-one was interested in the ideas of on inexperienced young man. However, I had confidence in myself but yearned for more challenging work.

When I was a draughtsman at the British Aerial Transport Company in 1918, I was engaged on undercarriage design and found them simple affairs, and the FK. 27 incorporated an oil dashpot. I saw nothing difficult in this work so that when I joined A. V. Roe's at the age of 19 in 1920, l did not hesitate to say, 'I knew all about undercarriage design. '!

I read up books on hydraulics and gun recoil mechanisms and found the classic formula for determining the resistance of oil through an orifice. This was most important to ensure that the energy of an aeroplane landing was completely absorbed. I still have my notebooks going back to 1920 with the work on this subject. There was no-one else in the A. V. Roe drawing office that knew anything about such things and so I was regarded as the undercarriage expert.

At the age of 19 I designed the undercarriage for the first Cierva Auto-gyro and the Avro Aldershot, the first aeroplane to fly with a 1,000 h.p. engine – the Napier 'Cub'.

When I left A. V. Roe for the Gloster Company I arrived at the age of 23 with a reputation as a good undercarriage designer for I had read papers in London before the Institute of Aeronautical Engineers on this subject, which had been well reported in the technical press.

Such were the steps taken early in my life that led to my later success.

I was restless for nothing gave the scope for exercising ingenuity as did aircraft work. The aircraft industry was wide open to new ideas whereas the old established companies refused to countenance departure from their traditional ways. Unfortunately, there were no vacancies in the aircraft industry at that time.

I designed a print drying machine for the photographic trade and started a business in Worcester. I lived with my mother in Sansome Walk, Worcester and later I moved with her to Oakfield House, Ombersley. It was here that my twin brother and I bought an old motorcycle which required much pushing to get it started and this unfortunately, over-taxed his heart and he contracted rheumatic fever, an illness from which he suffered for the rest of his life.

The business I started in Worcester did not produce many orders and I was soon looking for employment once more.

In 1921 when my home was at Ombersley I fell in love with a charming girl from Helensburgh – Arden McKenzie. She was staying with friends in that Worcestershire village but I had taken a job with A. V. Roe's in Southampton and in those days we worked a six day week. So there was no chance to visit Scotland and in any case I could not afford the fare. Thus my first love affair petered out.

My second eldest brother, Joseph, was now a chemist in Southampton and suggested I should come and stay with him, saying that A. V. Roe's of Hamble were looking for technical men.

In November 1921 at the age of 20 I started as a draughtsman with that company as their undercarriage designer. The aircraft on which I worked included the 'Bison', a Naval deck-landing aircraft, the 'Aldershot', the first aeroplane to be fitted with a 1,000 hp. engine, the Napier 'Cub', and the prototype 'Cierva' Auto-gyro.

In those days there was no section dealing exclusively with the strength of aircraft structures. In later years, of course, all such companies had stress offices. Each man did his own calculations and the first time I was expected not only to make undercarriage designs and the detail drawings but to calculate the loads on each member and ensure they were strong enough with an adequate safety factor. I am sure a young man of 20 today would never be given such a responsibility. However, I did slip up over an undercarriage for the dock landing 'Bison'. This had a large wheel track and a correspondingly long axle which carried, hanging down from its centre, a device for catching the ropes that stretched longitudinally along the deck. These were used for holding the aeroplane to the deck once it had touched down and, by friction of the ropes to the device, braked the forward run of the aeroplane.

As I have said, the axle was a fairly long one and being made as light as possible I had overlooked one important calculation – the deflection of the axle at its centre. Preliminary flights were carried out on the grass aerodrome at Hamble and as it was hay making time it seemed I had designed the best hay rake, but quite unintentionally!

The drawing office was small, consisting of about fourteen men. Later I employed many of them, Rowland Bound, V. O. Levick, Barney Duncan, Charles Blazdell, Clifford Tinson and Frank Bastow, all first-class men.

Roy Chadwick, their chief designer, had certain sayings which amused us. 'No-one takes any notice of me', and 'I distinctly remember telling you'. Neither of these statements were correct but were mannerisms. He was a good designer we all respected and, like others of that era, a great one for detail. I shall refer later to our close association when he placed considerable business with me.

During my lunch hour it was my practice to take a golf club on the aerodrome and hit balls against the hangar door, which then rebounded back. One day I hit a perfect shot which disappeared through a gap in the doors little bigger than the size of a golf ball. It came to rest in the wing of an Aldershot and I was most unpopular. !

On Sunday mornings A. V. Roe (Sir Alliott as he became) organised competitions for gliders launched from the factory tower. Such was local interest that attendances at divine worship declined dramatically, and the rector, the Reverend E. V. Roe (A. V. Roe's brother) intervened to halt further competitions.

One of the flying models I made in those days showed ingenuity.

I wanted to make a scale model flying boat but realised that it would be impossible to carve a hull and hollow it out to the required thinness, so I bought a bar of Sunlight soap, sold at that time in long square section bars, and carved the shape of a flying boat hull. I then coated this with layer after layer of silver dope, a type of celluloid. I continued with layer after layer of dope until I had a good thick skin, then I put the complete hull into boiling water to dissolve the soap which left a very robust hull to which I attached the wings. This model was extremely light and much admired.

In 1922 during my employment with A. V. Roe's I read a paper in London on 'Oleo undercarriages' before the Institute of Aeronautical Engineers. Although only 21 I was able to make this contribution on what was then a little known subject.

During my stay with A. V. Roe I knew their founder Sir Alliott Verdon Roe well, but his ideas for walking on the water and a new monetary system did not match his aeronautical achievements. Their test pilot, Bert Hinkler, was a good friend of mine. He was a great character with whom I often flew but he lost his life on a flight to his native Australia.

Life in Hamble was pleasant. I knew a Henry Molesworth – author of that well known engineers' handbook. He had an attractive daughter who was an olympic swimmer and a yacht called the 'Black Rose' on which he invited me for a cruise one weekend. With a good staff on board we were moving down Southampton Water in what to me was the height of luxury when he turned to me and said, "You know Dowty, I am a comparatively poor man". I have pondered that statement ever since.

In order to appreciate those problems that stimulated me in the development of landing gears it is necessary to recall the stage of evolution which the aeroplane had reached in the 1920s. As a flying machine it had vastly improved, both in reliability and performance under the stimulus of wartime development. The first inter-continental flights proved this and so did the spectacular increase in speed, for in 1921 the British speed record was just short of 200 miles per hour, which seemed quite fabulous then. With the increase in the aeroplane's performance it became apparent to me that in landing gear design there had been practically no progress at all. Landing wheels were still attached to a fixed undercarriage and were unbraked, while the tail was equipped with a skid. The aeroplane landed in a tail up position and it was not until the speed had fallen that the tail was allowed to drop whereupon the skid ploughed through the grass of the landing field to produce the only braking available. Once the aeroplane had touched down the machine became virtually uncontrollable, in as much as the pilot could neither change its speed nor alter its course. This primitive state of affairs seemed to be acceptable when landing speeds were low but with the increase in performance improvements became pressing.

It was in the mid-twenties that I gave lectures before the aeronautical societies on aircraft alighting and arresting means, and caused me to concentrate my activities on the development of improved undercarriages.

While I worked at Hamble I had lodgings in Southampton. This was the time radio started with those well-known programmes, 'The Savoy Orpheans', and 'The Savoy Havana Band'. All young men were avid readers of radio journals and the builders of wireless sets using accumulators and Bright emitter valves. Somehow I succeeded in setting my bedroom on fire so that I had to find other lodgings.

On leaving A. V. Roe in 1924 I was offered two jobs, one with Handley Page at Cricklewood and the other with the Gloucestershire Aircraft Company, as it was then called, at Cheltenham. I took the Handley Page job in May at a salary of £5 5s. per week but after one week I gave in my notice for I did not like the barn-like office or its atmosphere. I joined the Cheltenham company of which Mr. A. W. Martyn was chairman (he later became chairman of my company). On the board were the late David Longden and Hugh Burroughes, with the late H. P. Folland, chief designer.

The manufacture of aircraft was carried out on the same premises as that of H. H. Martyns Ltd., Monumental Masons. I found it extraordinary for sister companies to manufacture aircraft in one shop and memorials in another! Whatever happened there was always good business.

When moving to Cheltenham in 1924 I found since my first job in London in 1918 I had been in no less than 20 lodgings – or an average one move every 3½ months during those 6 years. The cost had varied between 20s. and 50s. a week. The latter figure was that charged in the Birmingham area and the lowest costs were in and around London.

When working with the Gloucestershire Aircraft Company I had lodgings in Byron Road, Cheltenham and from my rooms I could look across the road and see a young man, Stanley Evans – also employed in a technical capacity with the Gloucester company – either eating his breakfast or working on some papers.

Here was a smart young man, well turned out and with original ideas. Unfortunately these never materialised for he lacked the ability to apply them. Nevertheless, he wrote brilliantly with a fine command of the English language and his technical articles were excellent. His writings gained him recognition and positions which he was not able to fulfil. For instance, I was helpful in his gaining an appointment as chief designer of Folland Aircraft – a position he held only for a short time. This inability to hold a senior position in the aviation industry undoubtedly soured him, for he had an unfortunate way of insulting his contemporaries who got on. He finally went to California where he married and raised a family but he continued a long correspondence with his friends. But in the end we had to give up writing to him for he became such a cantankerous and bitter person.

On joining the Gloucestershire Aircraft Company I worked at their Sunningend Works in Cheltenham but later the design and engineering departments were moved to a country location, Brockworth Aerodrome. It was at that time I bought myself a secondhand Austin 7 car for £40 to provide transportation. This was the first car I had owned. Before then I had motorcycles. During my sojourn with the Gloucestershire Company I worked on many aircraft including the 'Grebe', the 'Gamecock', the 'Gloster IV' (the Schneider Trophy machine), and the 'Gauntlet'. I took out several patents for aircraft deck landing apparatus, disc wheels and shock absorbers. My relationship with the company over patents was interesting. In September 1925 I assigned my patent on deck landing arresting devices to the company for the sum of £150 and a royalty of not less than 5%. As this represented nine months' salary I was well pleased. In June 1927 I received a further £100 for full and complete rights to apply the invention to any aircraft manufactured or designed by the company. This scheme was never developed, it was a few years before its time.

Mr. Folland wrote me on 20th March 1928 saying that the company should be given the first opportunity on my ideas. I spent nearly all my spare time on a drawing board in my lodgings developing new projects.

I remember taking dancing lessons at the Irving Academy – run by Mrs. Irving, the mother of Charles Irving who later became Mayor of Cheltenham and my Public Relations Officer, and in 1974 the Conservative Member of Parliament for Cheltenham.

In October 1926 I gave a further lecture before the Institute of Aeronautical Engineers in London on "'Aircraft Alighting and Arresting Mechanisms'. This made headline in the daily newspapers. For many years I contributed articles to the technical press – *The Aeroplane, Flight, Aircraft Engineering, The Automobile Engineer, 'Les Ailes', India Rubber Journal, The Draughtsman*, and a thesis entitled 'Balanced and Servo Control Surfaces' was published by the National Advisory Committee as Technical Memorandum No. 563. All these efforts brought me additional income which was welcome and made my name known in the industrial world.

In 1926 I was concerned by the elementary way in which the blades of the Auto-gyro were started by the use of a rope. I put forward an idea for rotating the blades by turning the engine exhaust gasses through the tips of the blades. I still have the calculations to show such an idea was possible. It was not put into practice then but many years later the Fairey Company came out with a design for jet powering the blades.

The Institution of Aeronautical Engineers

A Society for the encouragement, development and protection of the Profession of Aeronautical Engineering.

This is to Certify that

George Herbert Dowty

was admitted _____ *Member_____ of the Institution

of Aeronautical Engineers on *28 September 1926* having

satisfied the Council that he had the necessary qualifications.

Admitted Founder Associate Member 30th October 1919

_____ President

_____ Member of Council

_____ Secretary   Diploma Number 7

34, Broadway, Westminster, London, England.

That same year I visited Paris for the first time to attend the Air Show. I was interested to see what I believe was the first aircraft wheel with internal springing. This was of Breguet design. The wheel travel was small with no provision for brakes. A somewhat similar wheel was used by the De Havilland Company on their record breaking 'Tiger Moth'. In 1927 I took out a patent for a wheel incorporating oil shock absorbers, steel springs and brakes operating on the wheel rim.

In November 1927 I visited the Air Ministry and met a Mr. Spencer Wigley to discuss the internally sprung wheel. A month later he wrote objecting to my proposals and in May 1928 I received a letter from Mr. Folland saying that the Gloster Aircraft Company had decided not to proceed with this development and gave me a free hand in all respects as to patenting. In 1930 I patented an improved wheel with the brakes imbedded in the wheel side and this was the design that was to set me up in business. I approached the Air Ministry again and after some delay an order was placed with the Gloster company for three experimental wheels which were sent to the Royal Aircraft Establishment at Farnborough for testing. The wheels came through the test successfully but this turned out to be a technical triumph only, it was no more. Mr. Folland was unable to offer any hope that his company could pursue this work due to lack of government support.

On May 28th, 1928 I entered into an agreement with the Dunlop Rubber Company, whereby they paid me £100 in respect of a patent on aircraft disc landing wheels. In October I wrote to the Dunlop Company saying in accordance with our agreement I was submitting detail calculations on my internally sprung wheel.

In 1929 I was offered a position with the Douglas Aircraft Company of California but my departure was delayed by quota difficulties. Subsequently they withdrew their offer saying they had a full complement of designers.

In 1928 I had applied for a patent on an improved springing and shock absorber strut. Hitherto these struts were bulky and being located in the propeller slip stream were the cause of considerable aero-dynamic resistance. As the struts were lengthy I used this feature to share the load between the two sets of springs mounted one above the other and operating in parallel. This reduced the width of the struts considerably and proved an elegant design. These struts were later used on the Gloster 'Gauntlet' and Saunders Roe 'Cloud'.

In an article at this time I remember saying that the ideal

undercarriage leg should be of no greater size than of the finest strut required to carry the maximum landing load. However, my efforts to induce aircraft constructors to take up this product were of little avail and my next line of attack was to try and float a company to manufacture these struts. I was convinced of the commercial future for specialised aircraft components manufactured by an independent company. Apart from commercial considerations the technical improvement of such components could be far more intensively and successfully pursued by a specialist manufacturer supplying components to the whole industry than by one aircraft constructor producing a limited number for his own use.

'Aircraft companies', I wrote at that time, 'do not favour the use of products manufactured by their competitors and an aircraft equipment company will be successful only if independent of aircraft constructors.'

Nevertheless, at that time I could find no-one interested in backing such a business. I spent money I could little afford patenting my novel undercarriage strut but it seemed I might as well have poured this money down the drain. I corresponded with many people – a Mr. M. L. Bramson, Col. the Master of Sempill, Mr. Gordon Smith, Merriams Aviation Bureau, R. H. Mayo.

In the years 1928-1930 I produced many technical reports, setting out the advantages of my proposed aircraft products, new types of undercarriage shock absorber struts and aircraft wheels with internal springing.

From 1930 onwards I put out well reasoned proposals for the formation of a company to manufacture and sell these products. I completed detailed drawings and calculated with accuracy weight estimates and obtained quotations for the manufacture of the parts to show my proposals were viable.

I found aircraft constructors somewhat incredulous at the prices I quoted. In 1939 when in production with the Gladiator internally sprung wheels I found the cost of the wheel with tyre, brakes, shock absorber and springing was £69 6s. which included my labour and overheads of £5 12s. I was selling each wheel for £104. These costs fully justified the predictions I had made ten years earlier.

Looking back at my early proposals it is difficult to believe that no-one was sufficiently interested to look at my designs, calculations and estimates for the claims I made in those early days all came to be realised.

George in the late 1920s studying a new design

When I wanted to start a business in the years 1929/30 these were days of the great depression and my estimates for what I required were extremely modest. I was seeking an investment of only £5,000 but no one was interested.

I was desperate for financial help. I wrote to the Bendix Company of America who acknowledged the letter but I heard nothing further. The Comper Aircraft Company said they could not make a decision about an investment until after the Paris Air Show. Merriams said they had placed my proposals before their clients but nothing transpired. I approached the Midland Bank but they refused me a loan. I asked if they would lend me money on my two life insurance policies but even that was refused. I approached local solicitors, Winterbotham and Gurney, who said they had placed my proposal before their clients but nothing came of this.

At any other time such a product might have found enthusiastic backers but in 1930 Britain was heading rapidly for a great depression and the aircraft industry was in the doldrums. Nobody had any money to risk on the development of new ideas, however promising.

There were already certain specialist suppliers of aviation equipment, Messier in France and Bendy in the U.S.A., who were independent of the airframe manufacturers and I saw the need for such a company in the U.K.

In 1931 only twenty-eight years had passed since Wilbur and Orville Wright had made their historic powered flight in Kittyhawk. Although development during the First World War had been rapid and designs had made remarkable strides, yet aircraft constructors seemed to build everything except the engine, the landing wheels and a few dashboard instruments. I judged there were opportunities for companies to specialise in the design and manufacture of specialized aircraft products.

I decided to set up a company with an address from which I could solicit business. It happened that my sister knew a Mr. Melville who had an office in the City of London and it was agreed that this office would be used as an accommodation address. In January 1931, Aircraft Components Company, 4 Lloyds Avenue, was formed. This was a ghost company with no capital, no staff and no premises. Because I was still employed by the Gloster Aircraft Company I could not use my name in connection with this venture. Letters received at Lloyds Avenue were forwarded by express post for my attention. Files from that period reveal the formidable amount of correspondence dealt with. All this work was done in my spare time.

Without capital or production facilities it is not possible to undertake any manufacture and subcontracting was the only way, but this is not a bad thing for the product costs are known and, furthermore, it is possible to shop around to find the most economic suppliers.

With the aid of my twin brother, the late Edward Dowty, a four-page illustrated brochure was produced announcing a new type of aircraft shock absorber strut. This, like the cost of stationery, was paid for out of my salary of £5 a week. Money was not available for advertising so articles in the technical press were the only means for publicising this product. The first order came from the Civilian Aircraft Company of Hull. Ironically this firm went bankrupt before they had paid for the struts but I was rewarded by an order from the Cierva Auto Gyro Company.

The parts for these struts were machined on a foot operated lathe by a Mr. Bowstead, who was to be one of my first employees, in the cellar of his house. It is difficult to understand how these units ever came to be flown for they were not covered by Aeronautical Inspection Department release notes – in fact I probably did not know such notes existed and apparently nobody cared. Those were happy days. I could use limitless possibilities and the prospect of developing a business looked exciting.

I had known for some years Joe Wright, who represented Dunlop's aircraft wheel and brake business and later became director of their aircraft interests. In 1930 I visited Dunlop for technical discussions and was sent to their rim and wheel factory in Coventry by Dunlop car. Being a very cold day I was kindly provided with a Dunlop hot water bottle which I placed at my back so that when I got out of the car my back suffered a sudden change of temperature. The following day I was bent double and in considerable pain with lumbar neuritis. In bed for two weeks the doctor's injections had no effect. Unable to remain idle I wrote an article with illustrations describing the internally sprung wheel, which was published in *Aircraft Engineering* for May 1931. I was a good draughtsman as can be seen by the drawings made at that time and the illustration in the articles of *Aircraft Engineering* obviously interested prospective users.

In June 1931 I offered Dunlop's an option to acquire my patents but they said they were only prepared to quote for the manufacture of the wheel shells. This was a lucky decision for that same day I came home in the evening to find this telegram:

'Referring article in *Aircraft Engineering* May issue, kindly telegraph name maker new type air wheel Figure 10. – Kawasaki'

It was fortunate that earlier that year I had set up a company so when this enquiry for four wheels was received I was able to quote from a business address. With no factory I obtained a price for the manufacture from the Gloster Aircraft Company (as it was then named) for £400 a pair. Kawasaki moved fast, for on June 18th they placed an order, signed by a Mr. Stanley Grave, their London office manager. When I placed the order with Gloster Aircraft they said the price they quoted had been incorrect and the price must be doubled. In desperation I left Glosters in June 1931 to manufacture the wheels myself, with no idea how I was going to get these wheels produced but I felt sure I could do so.

Starting in Business

At that time, Glosters, like other companies, were getting rid of their staff and I know my resignation must have seemed to my colleagues like a foolhardy action. Having secured this order I thought I could now find someone to back me and I spent three fruitless weeks approaching many people. I have a letter addressed to the manager of the Midland Bank, Cheltenham, asking what he could lend me against two life assurance policies of £100 each. As I appear to have had no reply I can only assume nothing came of this for I had to resort to money lenders charging an exorbitant rate of interest. I found friends in some Midland machine shops – firms who I had come to know through the buyer at Gloster Aircraft and they agreed, as did the Dunlop company, to give me credit.

I rented a mews loft at 10 Lansdown Terrace, Cheltenham, at half-a-crown a week and the only equipment I had was a work bench with a hand operated pillar purchased for £3 15s. and a piece of plate glass to serve as a surface table. The pillar drill is now in the exhibition room of the Dowty Group.

I worked night and day making production drawings, ordering materials and placing orders for the machine fittings. I employed two men who worked in the evenings assembling the wheels and they agreed to be paid when I was paid.

Melville insisted the premises should be insured against fire but I was aware that the premiums would be too great, for the ground floor was occupied by a wheelwright and covered in wood shavings, whilst next door was a garage and a petrol store. It is interesting to reflect that while one craftsman plied his trade with chisel and spoke on the ground floor a new kind of wheelwright was at work just above his head, using very different tools and materials.

It seems inconceivable now that such relatively complex units could have been produced under such circumstances in the space of nine weeks. Yet they were built on time and on the 21st of September the wheels left Cheltenham by rail. On the 25th I received a welcome telegram from Mr. Melville, 'Kawasaki paid, everything O. K. ' and later that month the steamer 'Yosukuni Maru' sailed down the Thames with the wheels safely on board.

Having left the Gloster Aircraft Company there was no longer any need for me to hide my light under a bushel. More stationery was printed with my name at the head.

During the time the Kawasaki wheels were being manufactured I had an enquiry from the Airspeed Company at York, of which Nevil Shute Norway was the managing director. He was building an aeroplane called the Airspeed 'Ferry' for Alan Cobham (later Sir Alan) and amongst other features there were two doors for the passengers, one for them to enter and another for them to get out, so that there was no delay in changing over the fare paying customers. This introduction with Nevil Shute Norway had come about by an introduction I was given by Mr. C. G. Cray, the well known editor of 'The Aeroplane' at that lime. In later years Norway emigrated to Australia where he became a very popular author writing under the name of Nevil Shute. We became great friends and he used me as one of his characters in his book "Trustee from the Toolroom". It is humorous to recall that when we first exchanged letters we were under the impression that the other's business was far larger and more soundly based than was the case.

This order necessitated a move to larger premises. It would be difficult to describe the company's first home at Grosvenor Place South as a factory, for it was part of the premises of a monumental mason. The adjoining yard was filled with samples of his melancholy craft. In fleeting moments of frustration I must have wondered whether the stones were to be symbols of failure.

The Dowty organisation really began in November 1931 when I engaged my first two full employees, John Dexter and Joe Bowstead. Both were employees of the Gloster Aircraft Company and that is how I first knew them. They were first class workmen to whom I owed much. Dexter combined the duties of turner, fitter, maintenance man, as well as doing estimating and planning. It was he that bought my first machine tool, erected the overhead shafting and belts. The workshop consisted of a bench, one centre lathe costing £25 and a small office where everything, designing, typing of letters, preparation of estimates, book keeping and the interviewing of visitors went on. The premises were heated by a coke stove. The salary of £5 a week which I drew at that time was all I could afford.

My early employees selected what machines were within my financial reach, and the only transport I had was my Austin 7, on which I carried out all the repairs and maintenance for I could not consider garage charges. I decarbonised the engine, ground in the valves, relined the brakes and did all the other minor repairs, and this efficient little car carried me all over the country for many years.

I was still unable to obtain financial backing and I doubt if I could have remained in business had it not been for Kawasaki paying me £2,000 for a licence to manufacture my wheels in Japan. This money enabled me to weather the next few years, for the Airspeed 'Ferry' undercarriages were unprofitable.

It is interesting to reflect on the wages I paid in those days. My first draughtsman received £3 10s. for a 40½-hour week, a first class machinist received 1/6 (7 ½p) an hour for a 47-hour week.

Incidentally, in the letter engaging this machinist I said l could only guarantee him employment for two months. In the event he worked for me for 40 years. At that same time I engaged a young man on a two weeks' trial at 10s. a week. Such were the rates acceptable in those days.

There were some things I had learnt from working in drawing offices. One was how attain the greatest efficiency.

Some offices were like large barns, with 50 to 100 men and the disciplinary functions of the chief draughtsman carried out by looking over his underlings from a glass box. Rarely did the chief designer visit the drawing boards.

There were other offices which used vertical drawing boards behind which draughtsmen could hide. That was a convenient way to waste time chatting or even sleeping. A man could leave his board and visit the toilet with a daily newspaper for long periods and never be missed. For myself I always want to look over a drawing office and see the measure of activity. I dislike vertical drawing boards.

There were smaller offices where the chief designer visited every board twice a day. What a difference. Men worked quickly and the drawing output was five times that of the larger offices. I know, for I worked in both.

In January 1932 I received a visit from a Mr. Doi of the Kawasaki Company who has given me the notes he made at that time. He came to collect data concerning the design and manufacture of the wheels we had supplied him . It took him 49 days that time to travel from Tokyo to London but when he came again in 1971 he did it in 16 hours.

By the end of 1931 shock absorber struts had been applied to Saunders Roe of Cowes, to the General Aircraft Company of Feltham and to Air Speeds of York. Nevertheless, the financial situation was still precarious. I still had to win what could be described as a production order.

In all this business I had to do practically everything myself. Type the correspondence on a portable typewriter, make my designs and working drawings, produce sales literature and quotations, visit aircraft companies to sell my products and, when orders arrived, place sub-contracts and undertake progress work. I drew up my own patent specifications and prosecuted them through the Patent Office, for I could not afford a patent agent. In this and other ways I kept my overheads down to an absolute minimum.

There were many disappointments. The decision of Roy Chadwick to use the internally sprung wheel on a new machine was very heartening, but after months of correspondence he wrote that his company had been compelled to discontinue development on this machine because of worldwide conditions. Everywhere it was the same story.

For some time it had been clear to me that the fixed undercarriage was only a half-way house, and that aircraft of the future would have retractable undercarriages. The idea of retraction was by no means new, but it had not been used on a production basis, one reason being that no lightweight power means were available for lifting the undercarriage. The economic depression, too, was holding back development. My business venture began with four dreadful years during which I tried to live off an industry that did not seem to exist. The stagnation of the aircraft industry was frustrating, and more than two years were to pass before a machine with a retractable Dowty undercarriage took to the air.

During those lean years measures had to be taken to keep the frail company alive.

In collaboration with Edward, my twin, we produced a photographic print glazing press and offered it for sale at £10. Another desperate effort to raise money was by the manufacture of metal garden labels.

In 1932 Aircraft Components Company became Aircraft Components Ltd registered as a private limited liability company with a capital of £1,000. The annual turnover then was a mere £2,800 and there were five employees. Twelve months later the turnover was £5,000 and the number of staff had risen to eleven.

In November 1933 we were supplying equipment to many aircraft types – the Saunders Roe 'Spartan Cruiser', 'Cutty Sark' and 'Cloud', the Phillips and Powis 'Miles Hawk', the Percival 'Gull', the Comper 'Mouse', the Gloster 'SS 19B', the Hawker 'Fury', and the 'F37' with internally sprung wheels, the De Havilland 'Gypsy', 'Leopard' and 'Puss' Moths, the 'DH84' and the 'DH86'. The list reveals the beginnings of a considerable business with exports not confined to Japan. We were selling to Siam, the Breda Company of Italy, Henschels of Germany, two firms in Sweden, and the Royal Danish Airforce. It would have been a different story if the number of components supplied to each customer had been considerable, but not a single one was a production order. Nevertheless, the way in which a back-street engineering shop with little capital had succeeded in two years in getting its products accepted by aircraft manufacturers in Britain and in much of the world was impressive.

I remember an interesting experience with Saro 'Cloud', an amphibious aeroplane with telescopic shock absorbers made from thin gauge stainless steel tubes. We found it impossible to prevent tubes from binding until we finally reached a solution with use of a certain brand of toothpaste!

It was in 1933 that I had my first business dealings with the Germans. The technical papers were frequently referring to my aircraft products and I received a letter from Herr Faudi of Frankfurt asking me to receive a Herr Brauer. As a result they took my shock absorber designs and we did business until the war broke out. At that time the Germans were not allowed to bring money out of their country, but come out it did, in the German's socks.

I went to Germany fairly frequently, staying at Kronberg in the Taunus mountains with Herr Faudi. Frau Faudi kept hawks and her sport was to let a rabbit out on her lawn and watch the hawks kill it. I was taken to Heidelberg to see a sausage skin factory. It was here I first became acquainted with Buna synthetic rubber – which resulted in my setting up a factory for producing synthetic rubber seals. I went to Lake Constance to meet Herr Eckner, captain of the Graf Zeppelin , who showed me over that great airship.

These German visits were most helpful. It was there I saw castings in magnesium, at the Elektronmetall company of Canstatt, that could be completely deformed without cracking. This resulted in my using cast magnesium forks for tail wheel units with great success. I never went anywhere without learning something and acting on what I had seen.

Many years later, in 1962, Herr Brauer showed me a report they had asked for in 1933 about the status of my company. It said the business employed 80 to 100 people, that I was life and soul and was considered reliable and honest! It recorded that I kept a stock valued at £300 and could be allowed credit of £100. In fact in 1931 I employed three people but by 1932 it was five and by 1933 it had risen to eleven.

About this time references in the technical press brought us more business with overseas companies who sought agency agreements. We had arrangements with Coghlin in Canada, Bjarne Sjong in Norway (who 40 years later were still doing excellent business for us) Finland, Romania and Poland. In 1933 I had my first enquiry from Fokker of Holland who, war years excepted, have been our good customers ever since.

Nevertheless, in November 1933 I was still in financial difficulties, badly needing money for development of my company. I wrote to a Mr. Ormonde Darby of Kingsway House, Kingsway, London offering him a 50% interest in my business for a paltry £5,000!

My turnover at that time was less than £5,000 a year but I had already supplied goods or had orders on hand from 19 British aircraft manufacturers including Gloster Aircraft, Hawker Aircraft, the Cierva Auto Gyro Company, A. V. Roe and companies in Italy Sweden, Japan, Siam, Denmark and Germany.

It is amazing how at that early period so many companies were interested in my products and yet no-one seemed anxious to invest in a company obviously showing great promise.

But this was the time of the great depression – the aftermath of the Wall Street crash. How fortunate I was that Mr. Darby did not take up my offer!

In December 1934, I offered an interest to Mr G. E. Beharrell, managing director of the Dunlop Rubber Company, explaining that in the last 14 weeks my turnover had been £6,900. Again there was no result .

However the same year I learned from my friend Mr. Folland that the Gloster Aircraft Company was expecting an order for the Gloster 'Gauntlet' biplane for the RAF. At that time Glosters were using Vickers oleo struts but because of financial restrictions they could not afford to experiment with alternative units. I could not afford it either but I knew my struts were better than those in use so I designed and built a pair for the 'Gauntlet' and gave them to Glosters, trusting their sheer merit would result in a production order.

These struts had steel springs mounted one above the other in parallel, and incorporated a damping device consisting of tapered cones which expanded under spring pressure and acted as a brake between the inner and outer telescopic tubes. The superiority of these struts won the day and they were fitted on the 25 'Gauntlets' ordered.

This was the first production order I received and marked the beginnings of a long association between myself and my old employer, the Gloster Aircraft Company.

In 1936 the 'Gladiator', the last of the biplane fighters, incorporated the internally sprung wheel, and similar wheels were fitted to the Westland 'Lysander'.

Another success came from the Bristol Aeroplane Company in 1934. I was driving back from Yeovil where I had visited Westlands, and reached a point where Bristol and Cheltenham roads forked. It had been a tiring day, and I was not at all sure it was worth paying a visit to the Bristol company. It was a good thing I did, for when I arrived I sought out my old friend Clifford Tinson, a designer there. I found him developing a tail wheel unit for the 'Bulldog' fighter.

In those days landing wheels with high pressure tyres were supported on a fixed undercarriage. When the aeroplane landed the only braking was through a tail skid which tore through the grass of the landing field. Once the wheels touched the ground the machine was virtually uncontrollable, as the pilot had no means of checking the speed or altering course. As aircraft speeds increased this primitive state of affairs was unacceptable, and gave me considerable scope for improvements in undercarriage design.

The 'Bulldog' had recently been fitted with wheel brakes and the tail skid, which till then had done the braking, was a positive handicap. Instead I designed and supplied the tail wheel units which allowed the wheel to caster freely on the ground but, when in flight, took up a true fore and aft position in relation to the aeroplane. I was delighted to get an order for 250 'Bulldog' tail wheel units for the Finnish government; it was this contract that really placed my company on a sound footing.

The first retractable undercarriage I produced was for the Bristol 142 called the 'Britain First', an aeroplane made by the Bristol Aerospace

Company for Lord Rothermere. He deplored the government's inertia over aircraft development. In those days the seals used in hydraulic systems were of leather and it seemed all the cows must have suffered from many pimples. Certainly we found it almost impossible to prevent leakage of hydraulic fluid and, furthermore, in those days we had no proper test equipment and out test beds were the airframes of our customers.

By the end of 1934 my staff had risen to 45 and we took new offices in Cheltenham's Bath Street. More workshop space was needed, and we took over a printer's premises in Grosvenor Place South.

It was not my company's ability to produce these assemblies that worried me, but the lack of adequate equipment for testing them. I had neither the money nor the space. Lack of capital hung around me like a millstone. From the original move to Grosvenor Place South the business had grown piecemeal, taking such buildings as I could find nearby, but the result was hideously inconvenient. I was tempted to sell out but at this critical time received encouragement from the late Mr. A. W. Martyn, the former chairman of H. H. Martyn and Co of Cheltenham and founder of the Gloster Aircraft Company.

Martyn guaranteed an overdraft with a debenture for £2,000 and my sister, Mrs. Fell, lent the company £100. In those days Mr. Martyn's home was 'The Hearne', Charlton Kings and I went almost every Sunday for tennis, dinner and snooker.

Our cramped quarters were still a great handicap and a new factory was the obvious answer. The Cheltenham authorities, however, frowned on building a factory on the outskirts of the town – an outrageous proposal not to be contemplated.

At this crisis time I heard of Arle Court, a mansion with 80 acres of well wooded grounds and eight cottages, three miles from the town centre. As an estate too large for a private house, it was a drag on the market, which is why I was able to buy it for £6, 500. The property was in the borough of Cheltenham and the authorities agreed to my taking it over, believing, I am sure, that I would be lost behind the trees.

Arle Court is a splendid building in gothic revival style. A great house with ornamental gardens, a lake, parklands and eight cottages. The house is magnificently panelled. There is a handsome staircase with carved figures – one of William the Conqueror and others of his descendants. It was probably from Arlette that the property was called Arle. [It is now believed to derive from the former name of the River Chelt.]

It seemed an odd choice for an engineering company, yet how well it turned out. The mansion became offices, the outbuildings, stables, coach houses and garage provided homes for men and machines.

The move could not have been more timely. The whole operation was carried out over one weekend and by Monday morning the machines were running again.

Now at last we had room to expand, and the aircraft industry was ripe for the contribution I knew I could make. Faced by the growing menace of Hitler's Luftwaffe the British government turned for help to the industry it had neglected for so long. The part my company played can be gauged from its growth. When we moved to Arle Court we had £5,000 of orders but four years later these had risen to £10 million.

On moving to Arle Court the capital of the company was increased to £30,000. Mr A. W. Martyn joined the board as chairman (a position he held until his death in 1947) and invested £5,000 for which he took a 15% interest. The Midland Bank lent me £10,000 against a debenture.

At this time I was supplying shock absorber legs to all the De Havilland Company aeroplanes. One day I received a visit from two of their senior men, Mr. Povey and Lee Murray, who came to complain about an invoice I had rendered for some trifling amount, but found themselves in an embarrassing position when they had to borrow money from me to get home!

There was a period in the mid 1930s when prototype aircraft were ordered from as many as three aircraft constructors. It was then my endeavour to get my undercarriages on all three prototypes so that whichever company won the order I was assured of business.

In 1935 I set up a company, V. P. Airscrews with Mr. Milner, the designer of the Hele Shaw propeller. Any company that had little or no profits pre-war was not permitted to make profits during the war and so this business was sold to Mr. Walter Jenkins, a friend of Mr. A. W. Martyn, a marble mason from Torquay who had built the Vimy Ridge Memorial and who was able to take advantage of the situation. Some eighteen months later Jenkins sold the business to Wilmot Breeden, who changed its name to Telehoist and later moved to more spacious quarters. It is interesting that the V. P. Airscrews factory came into the Dowty Group later on when it acquired the business of Gloster Engineering.

In 1936 I visited Stockholm for an aeronautical exhibition and met the three brothers Floorman. One was head of the Air Force, the other head of Air Force Procurement and the third chairman of Aero Material, our agents. Since then, the war years excluded, the Swedes have been our good customers.

At that time I made several visits to Holland and became a supplier of undercarriages to Fokker. We have always had excellent relations with that company and recently supplied important equipment for the F. 27 'Friendship' and the F. 28 'Fellowship' aeroplanes.

In 1936 I took out four patents of importance. The first was the levered suspension undercarriage which allowed the landing wheel to have a fore and aft movement and the shock absorber was firm jointed and relieved of all bending. This type of undercarriage was used on the Gloster 'Meteor'.

Another patent was the well-known aircraft hand pump – using a single cylinder but giving a full delivery on each stroke of the handle. This pump was standard on every wartime aircraft.

A further patent was the first undercarriage dashboard indicator and yet another patent was the particular use of torque links on an undercarriage shock absorber leg.

In 1938 I received a visit from Leonard Stace, a director of the Olley Group that had operated an air service to the Isle of Man and were involved in the building of Castletown Airport. Railway Air Services had taken over the Olley Company and Mr. Stace came to see if I could give him a job. It was then I suggested starting up a company producing synthetic rubber seals.

This had come about through that visit to Germany in 1938 where I had seen the manufacture of synthetic rubber products. Until then all our seals for hydraulic units – using mineral oil – had been made from leather, a most unsatisfactory material. Now it was possible to mould seals of high dimensional accuracy that would hold very high pressures and thus enable us to reduce the size and weight of hydraulic equipment – of great importance to the aircraft industry. There is no doubt that synthetic rubber put the hydraulics industry on its feet.

A company was set up under the name of M. E. Stace Limited, with a factory in Winchcombe Street, Cheltenham. This was success from the moment it started and supplied my company with synthetic rubber seals throughout the war.

Unfortunately this was another business without a profit standard prior to the war and however much we invested in plant and equipment we were not allowed to make a profit. Like V. P. Airscrews this business had to be sold to someone who could benefit from our initiative and enterprise. We found a purchaser in a Mr. Matusch, a paper doily manufacturer who could not get paper during wartime for his products but could profit from the rubber business we had started since he had a considerable pre-war profit standard!

After the war Matusch closed the business, thinking I suppose that orders would seriously decline, and went back to his factory in London.

It would be pleasant to record that all undercarriages I produced were trouble-free. I recall the flight of the first Hawker 'Fury' at the S. B. A. C.'s flying display at Hendon. The shock absorber legs were of the type which only reached their full extension by weight of the wheels and axle when the machine was airborne. Bill Pegg, the pilot, flew the aeroplane in an inverted manoeuvre, with the wheels and landing gear uppermost, so that when he executed a flick roll the wheels and axle were thrown outwards with terrific force. The anchorage to the airframe has never been designed to take such loads and fractured. To my horror I saw both landing wheels leave the aircraft. It was one of those moments I wished the ground might open up and swallow me. I thought this was the end. Fortunately the pilot was able to make a belly-landing with relatively small damage to the machine and none to himself. Nevertheless it was very alarming. I was sent for by Mr. Camm (as he then was), the chief designer of Hawkers, and you can imagine my feelings when ushered into his presence. It is always the right policy to admit one's mistakes and not make excuses. I obviously did the right thing in admitting the design fault which had not taken account of this unforeseen manoeuvre. It is happy to relate that from that day I never lost an order from the Hawker Company. Camm and I became firm friends.

Sir Sydney Camm was a man who took a lot of knowing. Even those who knew him well found in him characteristics which could only be described as peculiar. I believe most of this was due to an inherent shyness. He had a habit of ringing me up asking why I had not been to see him. Thinking it was a matter of some urgency I would arrange a visit within the next few days, yet arriving in his office I was somewhat taken aback by being asked for what purpose I had come! One had to listen to a tirade against civil servants and anyone he thought was getting in his way. He was really the friendliest of persons and would not only talk of business but of photography and golf which were his keen interests. On my leaving he would always come down to my car to see me away.

Under the Treaty of Versailles, Germany had been forbidden to build military aircraft and it was only in 1935 that the world realised that her commercial planes were in fact war planes in disguise. The German machines were much superior to any in the R.A.F. The aeroplane, whether it was a fighter or bomber, was now a low wing mono-plane with retractable undercarriage. To recover from fifteen years of neglect the British aircraft industry had to work fast. For me and my struggling company this was the moment of great opportunity.

In April of this year I engaged my first draughtsman, D. G. Bridges. Until then I had made all my own designs and detail drawings.

With the business still expanding I engaged Mr. R. H. Bound in 1935. My earlier work in the aircraft industry had brought me in contact with many men whose abilities I appreciated and who later joined me. That year Mr. R. F. Hunt, now my deputy chairman, joined that company as an apprentice coming from Cheltenham Grammar School. My salary at that time was £500 per annum.

From those early days I have taken the keenest interest in the education of young men. The apprentice system, now almost unknown in other countries, has served my companies well and I am proud of those men who, coming to us in their early days, have made good and reached top positions in Industry.

One of the first men I engaged on accountancy and book keeping was an odd character. At that time we worked six days a week, on Saturdays till noon. The first week this man was in my employ he asked to see me privately on Saturday morning. I thought maybe he had found the work not to his liking and wanted to give notice. He arrived in my office with a brown paper parcel under his arm and, proceeding to unwrap it, I could see it was an album. Here was a man undoubtedly going to show me some of his handiwork – but no – opening the album he said to my amazement, 'This is a photograph of my mother-in-law in her coffin'!

In those days I drove myself for I could not afford a chauffeur. One day this character, knowing I was motoring to London, asked for a lift. I thought it was strange that everyone was staring at my car. Turning round I saw my passenger wearing a tall astrakan hat and looking very inch an Eastern potentate!

During wartime he was the only employee to arrive at work on horseback. He also rode his horse into town shopping, tethering it wherever he could.

The Business Goes Public

In January 1936 additional finance was needed and the Midland Bank put up a further £5,000, and my sister £400. The Midland Bank told me this was the maximum borrowing they could permit and if further monies were required it would have to come from other sources. I realised too late that I should have approached other banks for help but I had no-one to advise me. I suppose I thought that if my own bank would not help then the others would also turn me down. Looking back at the success I had achieved it seems incredible how little confidence my bank had in this developing business and how badly they treated me.

As further monies were needed it was with great reluctance that a public company was floated. In March 1936 an issue of 800,000 shares of 5s. was made to purchase the company and provide the working capital that was needed.

Looking back at my assets – the sum paid for the business was pettifogging. For instance, Arle Court with its fine buildings, houses and 80 acres for development, although as a residence it was bought for £6, 500, its value as an industrial property was very considerably undervalued.

The prospectus then issued showed aircraft equipment was being supplied to Airspeed, A. V. Roe, Blackburn Aeroplane, Boulton Paul Aircraft, Bristol Aeroplane, British Aircraft Manufacturing Company, Cierva Auto Gyro, De Havilland, General Aircraft, Gloster Aircraft, Handley Page, Hawker Aircraft, Heston Aircraft, Parnell Aircraft, Phillips and Powis, Popjoy, Saunders Roe, Short Brothers, Armstrong Whitworth, Westland Aircraft and to the following aircraft operators – Imperial Airways, Air Work, British Airways, Hillman Airways and Jersey Airways.

Orders had been completed or were in hand for the following countries – Canada, South Africa, Belgium, Denmark, Germany, Holland, Hungary, Japan, Lithuania, Norway, Poland, Romania, Siam, Spain, Sweden, Switzerland and Yugoslavia.

There were eight published U.K. patents and twenty-five patent applications, and a further twenty foreign patents on file.

The man behind this flotation was Louis Jackson, whose girlfriend was the actress Enid Stamp Taylor. They often visited me in Cheltenham. His offices were at 50 Pall Mall which many years later, quite by chance, became the location of our London office. The brokers to the issue were E. R. Lewis whose head was Ted Lewis (now Sir Edward Lewis) who as Chairman of Decca Ltd. brought that company out of the doldrums to the great company it is today. Edgar Granville (now Lord Granville), then Liberal Member of Parliament for the Eye division of Suffolk, was brought onto our board by Louis Jackson. At the time he was parliamentary private secretary to Sir John Simon, the foreign secretary.

The forced selling of my company was a blow to me. What did I receive for a going business, plant, patents and the freehold Arle Court property? 222,000 shares of 5s. each, a salary of £1, 200 per annum and a commission of 10% on the net profits. If the commission had ever been paid it might have been some compensation for having parted with my business in its infancy but it never was, as I will explain later.

I have always felt aggrieved at the Midland Bank which, for the sake of a few thousand pounds, forced me to sell my business when two years later they had to lend me millions of pounds against government contracts.

After moving to Arle Court there was a building programme which never seemed to finish; before one building was completed another one was started. The work was undertaken by a local firm of builders but when I asked other firms to tender for a new building the local contractor said he was the only one who could quote so long as he was working on the site. I therefore told him to move his men and equipment from the site forthwith. I then engaged my own work force headed by a foreman who, from sketches I supplied, ordered materials, supervised the erection of steelwork, installed the drains, lighting and heating and, without other supervision, completed first class buildings with great economy, buildings that are in fine condition forty years later. The drawing office, which has a wall of glass bricks, a novelty at that time, was built at a cost, including lighting and heating, of 28s. (140p) per square foot and the factory for 65p per square foot.

With the considerable orders placed by the aircraft industry the Ministry of Aircraft Production become interested, perhaps I should say concerned, and sent a deputation to look at my facilities. This deputation was headed by Mr. Walter Pucky (later Sir Walter). They expressed the view that I would never cope with the volume of work on hand. I was annoyed with them at that time, but looking back I can understand their feelings. History records we did succeed in meeting our commitments and this was only possible with help I received, organising the business, and tribute must be paid to James Currie, a nationalised American who had come to live in England to take charge of Stevenson, Jordan and Harrison, industrial consultants. His assistance throughout the war years was of the utmost value and credit must be given to him for the paperwork he introduced about which I had little knowledge.

About this time I installed my first television set. Although the only broadcasting station was the Alexandra Palace, North London, yet with the help of the Pye Company of Cambridge and a very tall aerial, I was able to receive transmissions from 100 miles away although the quality was extremely poor by present day standards.

In 1937 Harry Folland became managing Director of Folland Aircraft of Hamble, the Gloster Aircraft Company having been taken over by the Hawker Siddeley Company.

The financial man behind Folland Aircraft was the late Alan Good. Mr. Folland asked me to put up a sum of money as a temporary loan pending the flotation of the company and this I did in view of my long friendship with him. Unfortunately, when the time came for repayment Alan Good invited me to his London office and said he would refund the money if I agreed to his acquiring Dowty Equipment! I told him what I thought of his business ethics. Mr. Folland was much annoyed over this matter and later repaid the loan himself.

In 1937 I received my first award for my technical work, the Royal Aeronautical Society's Edward Busk Memorial Prize.

It was while I was attending the Paris Air Show in December 1938 that I received a telephone call telling me of a motor accident in which my mother had been seriously injured. I left for home immediately in time to be with her before she died. Her passing was a great shock for she was an adorable mother beloved by all her family. Until 1930 she had lived for ten years in the Isle of Man at a delightful property called 'Thaneshurst' on the coast outside Laxey, where I spent many happy holidays as a young man. She could never see anything wrong in anything her children did and I remember her saying to someone, 'If one of my sons broke into a bank I would know it was for a good reason!'

Two years later I had the organ at Pershore Abbey rebuilt in memory of her and my father.

This was the time of many new designs recorded in the number of patents taken out covering retractable undercarriages, cockpit indicators, hydraulic pumps and controls, hydraulic packings and then in 1939 the first patent for a liquid spring shock absorber. These patents of mine were used throughout the war years and were instrumental in the rapid escalation of sales.

In 1938 there was a need to provide aeroplanes with a means for landing on beaches and very soft ground. It was then I proposed the use of a caterpillar track landing gear and the Ministry of Aircraft Production gave me a contract for the construction of such gear for the 'Lysander'.

When this work was in progress a visit was made by engineers from America and as a result the U.S. Air Corp built similar equipment for a small 'Fairchild' aeroplane. Following successful tests the Air Corp then asked for caterpillar tracks to be constructed and fitted to a 'Boston' of 19,000 all up weight.

The flight tests behaved well and the engineers of the landing gear division of Wright Field were well satisfied with the results when I visited them in June 1943. However, it was clear that an aeroplane fitted with these devices must be designed to permit retraction of this gear. With the development of the helicopter and now the vertical take off and landing aeroplanes, there is little point in pursuing this work as obviously it was more complex than the conventional pneumatically tyred wheel.

I had for long pondered the possibility of replacing the rather cumbersome shock absorbers hitherto used on aircraft. We had gone from tension rubber to compression rubber and steel springs and then to pneumatic legs of large diameter, and although we had always been told that liquids were incompressible I decided to carry out a simple experiment with a very small cylinder and piston rod which I closed in a bench vice. This seemed to show that liquids were compressible and further tests confirmed this. It was a case of trying to do something that was "not in the book". It resulted in the smallest size of shock absorber and spring unit ever designed and has been used on many aircraft throughout the world.

My first visit to the U.S.A. took place in the summer of 1939. I travelled on the Queen Mary and dined in the Verhanda Grill. One evening the ship's doctor gave a cocktail party attended by ten glamorous American show girls, some of whom I met later at 'The Diamond Horseshoe', a well-known place of entertainment run by the late Billy Rose.

As the flat coastline of Long Island hove into sight it was a clear and sunny day. We passed the Statue of Liberty and up the Hudson River, past countless enormous skyscrapers. A multitude of boats lined the shores of the Hudson River and numerous ferry boats and all kinds of craft moved hurriedly in every direction.

Shortly before docking in New York, immigration officers came aboard and when interviewed I was told to step to one side. I found myself segregated with a motley collection of unhealthy looking people and could not understand why I was not allowed to pass quickly through immigration like most other passengers. After some time I was presented to an immigration officer who having looked at my declaration proceeded to write out in triplicate, a certificate to the effect that as I had an artificial eye it was unlikely I would be able to earn a living in America! Over the years I have collected so many of these certificates I had almost enough to paper a room.

It is peculiar that every time I enter the States by boat I found the same ritual yet on arrival by plane no-one was concerned about my disability.

I have always found New York a most impersonal city, with streets at right angles to one another and gigantic buildings bearing down on one. It has not the charm or warmth of our old continental cities and I pity those who have to live and work under conditions which, to me, are singularly unhealthy with overheated offices and hotel rooms, and in the streets contaminated with gasoline fumes.

Unlike cities such as London and Paris which can be approached from all directions, New York being on a peninsula has much fewer approaches and congestion at busy times of the day is most frustrating.

When I disembarked in New York I was met by the late Sam Niedelman, one of the first graduates in aeronautical engineering from the Massachusetts Institute of Technology. He was operating a factoring business called 'Aviquipo' and had been our representative for some time. On my first night in New York he took me to see 'The Boys of Syracuse' and then twice to see 'Oklahoma', the first of that splendid run of American musicals.

He later told me he was expecting to meet a 'typical Englishman with spats and a monocle' but was surprised to find a go-ahead businessman in his late thirties. Sam and his charming wife Hilda became my very good friends and although Sam died some years ago I never visit New York without seeing Hilda.

Having now set up businesses in Washington D.C. and in Pittsburgh we no longer need the services of Aviquipo in the U.S.A. but they still represent us in South America.

On my first visit to the States I went in fear and trepidation having been told how far the Americans were ahead of U.K. industry. I called on many companies operating in my field of activities but was not impressed. None of them seemed to have original ideas and carried out work only as dictated by the airframe manufacturers. Their research and development teams were non-existent. In contrast to my company they had no proprietary products. The U.S.A. airframe companies bought bits and pieces from many different sources and strung them together, a hotch potch business which has continued throughout the years. As an airframe company produces a new aeroplane only every three or four years their experience on the design of equipment is limited compared to a company like mine which works on many types of aircraft simultaneously.

I considered the Americans lagged behind Britain in the design of aircraft landing gears and hydraulic equipment. The reason was that each aircraft constructor designed his own equipment so that there was multiplication of effort throughout the industry, a lack of standardisation and above all, lack of experience compared with the specialist manufacturer. Accessory manufacturers worked on aircraft constructors' own designs and so possessed little or no technical background and had no development teams.

When I brought the novel liquid spring shock absorber to the notice of Cleveland Pneumatic Corporation of Cleveland, U.S.A. , then the largest manufacturers of aircraft shock absorbers, their designer, the late Mr. Wallace, was reported as saying, 'That fellow Dowty is nuts, he doesn't realise it is the cylinder that is expanding.' Some years later in 1956 when our U. S. patents expired, that company manufactured liquid springs for American aircraft, notably the 104 Starfighter.

I called on principal aircraft manufacturers in California, Douglas and Lockheed, Ryan of San Diego, Boeing of Seattle and the Cessna Company of Wichita, Kansas. I visited the Golden Gate Exhibition in San Francisco, a really first-class exhibition, but I returned to the U.K. thoroughly heartened at what I could see were our progressive ways compared with our American counterparts.

George in San Diego, May 1939

In late 1938 the Canadian government made moves to expand the Canadian Aircraft Industry and it became obvious that we should take part in this. I visited Canada in 1939 and over one weekend arranged for the incorporation of Dowty Equipment of Canada. The first office was a room on the premises of B. J. Coghlin who had been our agents since 1933. Initially all manufacturing was carried cut by the Coghlin company.

Early in my business career I was concerned that there seemed no office desk with features that I considered most desirable. I was always spending money on up-to-date plant and felt it time for my office to be better equipped.

I designed a desk that had metal ducts leading to a water filled drawer to prevent smouldering cigarettes and unsightly ashtrays. Telephones and dictaphones were concealed, the blotter was sunk in the desk enabling drawings to be laid on the desk without obstruction. Most desks are cluttered with items that need not be there and a clear desk looks and is efficient.

The War Years

At the outbreak of war a German aircraft flew over Arle Court. I saw it quite clearly with its German markings. It came and went unmolested. After the war I was presented with photographs taken of my factory from this plane which are now in my possession and evidence that ours was a marked factory.

The Germans used radar beams which intersected over the area they proposed bombing. Means were found of determining the position of these beams and hence the intended target of the Germans. One day I was told in secret that the intersection of the beams was over Brockworth, the home of the Gloster Aircraft Company, only a few miles from our main factory. There was nothing to be done but to wait in fear of the outcome that evening. However, no raid took place on Brockworth that night or at any other time. I am told that these false alarms were not infrequent but we were not to know this.

During wartime there was need for much secrecy. M.I.5. were always in communication with me regarding certain people in my employ and the need to keep them under surveillance. I was requested to install a private telephone in my office that went neither through our telephone exchange or yet through my secretary.

One day I received what was obviously an important communication. It was delivered by special messenger and heavily sealed. It had to be delivered to me personally with instructions that only I was to open it.

Inside this large envelope was yet another heavily sealed envelope, saying it was to be placed in my safe and only to be opened when I was given specific instructions to do so. Here undoubtedly was a highly important document, nothing less I supposed than plans for the destruction of my plant on the arrival of the enemy. Some days went by and then I received instructions to immediately open the envelope. I went to the safe in fear and trepidation. Breaking the seals I opened the envelope to find a message from the Ministry of Aircraft Production to say they had moved their materials division to Shepton Mallet in

Somerset! What an anti-climax, what a joke when I had heard of this move some weeks earlier from the Ministry department concerned.

In 1939 my company was already heavily committed to almost every U.K. aeroplane but when War was declared our programme became so great that we employed the engineering resources of large companies such as I.C.I., Lever Brothers, Courtaulds, Metropolitan Vickers, Westinghouse Brake and Signal, and Gillette Stephens in addition to hundreds of smaller ones. The placing of contracts, the allocation of materials, the day to day liaison, the settling of innumerable queries, all these were controlled effectively from Arle Court, such that at the end of hostilities we could claim the unique record that not a single aeroplane throughout the war years had ever been grounded for lack of a Dowty spare.

Those of us who lived through wartime are not likely to forget the black-out. Automobile headlamps were restricted to an aperture the size of a penny. Travelling in fog was a nightmare which I found when motoring in Cheltenham one evening. Hugging the curb I took the wrong road and in reversing felled a tall lamp standard across my car. On another occasion when motoring to Liverpool, I felled a telegraph pole. On this occasion the location of the accident could not have been better contrived. It happened opposite a garage whose owner promptly sold me another car by cheque written on a sheet of notepaper.

In February 1941, we were honoured by a visit from His Royal Highness, the Duke of Kent, who shortly afterwards was killed in an R.A.F. plane when flying at night over Scotland.

This same year I introduced a feature into hydraulic rams which is used worldwide today – the use of back up rings on piston seals.

Aircraft Components was not a satisfactory name for my company. It was copied by others using prefixes such as 'Northern''. The words 'aircraft' and 'components' are names in common use and it was for this reason we changed our name to Dowty Equipment in October 1940.

In 1940 we acquired our first executive aeroplane, a single-engined Cessna at a cost of £400. We later bought a second one, not in a good state of repair, which we cannibalised to provide spares for the first aeroplane. Our pilot then was Martin Fountain Barber, our chief service engineer. This aeroplane was in great demand flying chiefly to Air Force stations and to distant sub-contractors, and it was in this aeroplane that I flew on the memorable trip over Cirencester and Tewkesbury when I spotted that forage store at Ashchurch railway station, about which I will say more later.

Fountain-Barber was the most accident-prone of anyone I have employed. When I arrived at the interview when I engaged him, he came blood bespattered having come off badly in an encounter with a lorry carrying bricks. He was hurt later in a collision with a tram car in Leeds, and then took off the wing tip of the Cessna aeroplane on a balloon barrage cable. On another occasion he was seriously hurt in an accident on Staverton Airport and spent several weeks in the Radcliffe Infirmary, Oxford. When due to come home I was returning from London by car and I called to bring him back. The car was loaded with his suitcases and it must have appeared that there was someone on a holiday jaunt. I received several visits from police who were hard to convince, for at that time there was strict petrol rationing and Ivor Novello had been imprisoned for misuse of petrol.

During the war years a room in my house was used as a plotting station for movement of enemy aircraft. Everyone had to undertake some additional wartime duties, either the Home Guard or fire watching. I chose the latter and the register was brought to my bed for signature.

I was continually aroused at night when raiders were observed to be coming in our direction, most of them destined for Liverpool, Birmingham or Coventry but there were occasions when bombs were dropped around my house and factory. On one occasion the lodge at my entrance gates was badly damaged and my office windows blown in.

I remember an amusing incident concerning the swans on the lake at Arle Court. Normally they were harmless and indeed wandered into the factories where the workers fed them. However, when their young arrived the male swan was vicious and would let no-one near. It would fly straight at any intruder. An employee of Stevenson, Jordan and Harrison, a Mr. Evans-Hemming, declared that all one had to do to avoid attack was to be brave enough to stand one's ground and stare at the on-coming swan.

One lunch time a large crowd gathered to see Mr. Hemming demonstrate his theory. He boldly stood his ground but the swan flew on unconcerned at his bravery knocking him clean off his feet to the great delight of onlookers.

In 1940 I submitted a design to the Air Ministry for a ground to air missile for the destruction of enemy aircraft. A patent was applied for in September of that year. This idea was obviously too advanced to be pursued at that time but after the war it provided a completely new industry. We took no part in these later developments. All this work was highly confidential and for one customer, the U.K. government, and I decided to pursue a development policy on products for worldwide sale without government restriction.

In 1941 I was suddenly stricken with a severe pain below my chest which my doctor diagnosed, to my surprise, as appendix trouble. Coming round from the operation I was told I had received double surgery. Finding my appendix ok they then looked around to find where the trouble really lay – a large gall stone!

During wartime I had of necessity to spend much time in London and often over night. I stayed at the Savoy, a hotel I have used for nearly forty years, for in all that time no matter what short notice I gave they have always found me accommodation. The riverside suites are splendid and quiet with views from Westminster Abbey to St. Paul's and the continual movement of river traffic is always a source of considerable interest.

In 1940 we sent two senior men to Canada to start manufacturing there in our own right. A workshop of 10,000 sq. ft was obtained and soon this was extended to 28,000 sq. ft with 8,000 sq. ft of office space. By 1942 there were 350 employees but most of the machined parts were bought from subcontractors. To give some idea of the volume of work carried out during war years our Canadian company supplied over 1,200 sets of 'Hurricane' equipment, 3,000 sets of Anson equipment and 700 sets for the 'Lancaster'. But this was not all, as products were produced for other aircraft types.

We set up a design team to produce ski landing gear for the 'Bolingbroke' aircraft, the first undercarriage for a major aircraft designed and manufactured in Canada. Eight service engineers were employed to cover 40 Air Force stations and the many manufacturing plants of subcontractors, which included such well-known companies as Cockshot Plough and International Harvester.

Soon after commencement of hostilities I received a telephone call from Tom Bata, that dynamic international manufacturer of boots and shoes, enquiring if I could find work for his modern machine shops at Batawa, Canada. These had been set up to make boot and shoe making machinery. This factory was well equipped with good management and throughout the War it was largely engaged on the manufacture of Dowty products. It was not until 1949 that I met Tom and his charming Swiss wife, Sonia, in Bermuda but since then we have been close friends.

The financial investment we made in Canada was on $1,000 and from its inception, throughout the war years, it was always self-supporting.

I found Canada and the USA had peculiar drinking laws. In the USA it was illegal to buy drinks between 3.00 and 5.00am but everybody seemed to know a place to go between those hours!

In Canada you could not be served with a drink whilst standing up and in Alberta no alcohol to be had in hotels. To get alcohol one had to get a licence from a government office and go to the liquor store and drink in the bedroom! In the Banff Springs hotel there was dancing but no drinking except tap water from plastic cups!

During the War I visited Canada and the USA each year. After America entered the war, air travel by US Air Force Command was easy but prior to that I had to take longer and more hazardous journeys by sea. In 1940 I left Liverpool on the S. S. 'Samaria' for Halifax. From letters written at that time the voyage was a nightmare. At all times we carried life jackets and were battened down in our cabins with all port holes sealed;there was no ventilation. The ship made 150 miles on most days in rough weather and the voyage took 14 days.

I returned on the 'Britannic' with Sir Roy Fedden who was a most depressing companion when we heard of the fall of France. He was of the opinion that all was lost. There was a mixed cargo aboard including, to my surprise, a large consignment of goldfish! I thought these must be destined for the desks of our civil servants to cheer them up but was later informed they were for reservoirs to detect the presence of poisons.

With the fall of France the Canadian companies were in difficulties due to supplies of materials from the parent company in England. I visited Lord Beaverbrook, the Minister of Aviation, a Canadian, for permission to send certain supplies but found him uncooperative, using his words, 'Not a damn nut or bolt must go to Canada'. I had pointed out that what we proposed to send would not in any way affect our production.

When I returned home I found an urgent message from Lord Beaverbrook asking me to see him. I retraced my steps to London and after the usual wait, all he said was, 'Good morning, will you let me have a list of the parts you want to send to Canada. Good morning'. With all Lord Beaverbrook's good qualities it was clear he was inconsiderate of a person's time. That was not the last time I was to receive similar treatment from a minister.

I continued to press Lord Beaverbrook on this matter and finally he telephoned me one day when I was at Westland Aircraft, Yeovil, to say he had appointed a Mr. Jack Bickell, another Canadian and a friend of Lord Thomson, to deal with this. A meeting was called at the Ministry of Aircraft at Millbank and although Jack Bickell was present he did not even sit at the table but stood at a window overlooking the Embankment and took no part in the discussion. I found this rather odd. Some months later when I could get no action I decided to ring Jack Bickell and to my surprise received a favourable decision the following day!

Nevertheless, Lord Beaverbrook was the only Minister of Aviation to keep in personal touch with me on any matters in which he thought I could be of help.

During the war, London was not exactly a restful or peaceful place. Air raid sirens, the arrival of bombers, the noise of anti-aircraft barrage and finally the doodle bugs. Fortunately I was never too near the scene of bombings but I lost many friends including James Ogden, the jeweller from St. James's Street, who I met on a visit to Switzerland in 1938. He was an air raid warden and went on duty one night but never came back. His body was never found.

The list of aircraft we equipped during the War, mostly landing gear and hydraulic equipment, were the 'Hurricane', 'Lancaster', 'Halifax', 'Typhoon', 'Lincoln', 'Whirlwind', 'Tornado', 'Blenheim', 'Stirling', 'York', 'Manchester', 'Hampden', 'Beaufort', 'Henley', 'Lysander', 'Anson', 'Rapide', 'Dominee', 'Tempest', 'Botha', 'Gladiator', 'Welkin', 'Skua', 'Hart', 'Flamingo', 'Shark', 'Fulmar', 'Barracuda'. A formidable list of aeroplanes. In January 1943 we had no less than 90,500 complete assemblies on order for spares alone, such as tail wheel units, pumps, etc, each involving some hundreds of items.

The Gloster 'Whittle', Britain's first jet propelled aircraft, was Dowty equipped and the famous Gloster 'Meteor' was the first retractable undercarriage in which the undercarriage leg shortened during retraction to permit their stowage in the small space available. This was an example of research and development being carried out without hindering wartime production.

To give some idea of the amount of equipment we supplied on some aeroplanes we estimated that one out of every sixteen 'Lancasters' was our responsibility. On this one aircraft over ten million man hours were absorbed. We produced 12,900 sets of 'Hurricane' equipment exclusive of spares. During the war we supplied 90,000 undercarriages and over one million hydraulic units. For some time during the war our production director was Mr. Wegerif, who joined our board in 1942. Other than myself, he is our longest serving director. He was loaned by Sir Charles Clore, owner of Gillette Stephens of Bookham. This company was our largest subcontractor during wartime.

An important service provided during wartime was the posting of engineers to aircraft constructors and RAF stations to ensure the correct installation of our equipment. At the peak of wartime activities my company had 3,000 employees of whom half were women. This labour force was backed by an army of many thousands of people employed by our subcontractors. These included 50 companies machining and assembling our products, another 106 companies carrying out machining parts only, 4 companies were producing hydraulic pumps, and there were over 150 companies supplying materials and small parts.

I was criticised for the amount of subcontracting undertaken. The reasons for subcontracting have been many. There has always been insufficient labour and I have had to send work to where labour was available. In the early days I did not have the capital to invest in machine tools so subcontracting released what money I had to get on with more projects and to devote my time to designing and marketing.

There is also the matter of work in progress. The subcontractor carries this cost which is an added advantage. There is also the ability to subcontract to specialist manufacturers, those for instance with multi-spindle automatics and others who produce exotic materials, and by machining these high cost metals can immediately return the swarf to the melting pot. In this way it is possible to buy these parts at less than the cost of the raw material.

The motor industry highlights the value of subcontracting. They go to specialist suppliers for all their components.

This dispersal of production had great advantages. It took the work to where the labour was available and it minimised the loss of capacity by aero attack. The subcontracting organisation was controlled by a staff of 600 which included engineers and inspectors who visited the subcontractors continually. This demanded a tremendous sustained feat of planning and co-ordination. Each Monday morning I took a production meeting dealing with every single contract and problem. These meetings started at 9.00 a. m. and finished around midnight. On the previous day, a Sunday, I would spend hours looking through the Stores and at incoming deliveries so that I was often better informed of the position of supplies than those who were in charge of this work.

One of the greatest concentrations of subcontractors was in Coventry and the blitz on that city was a most serious blow for much capacity was lost. I visited Coventry with the object of finding skilled men who had lost their jobs in order to persuade them to move to Cheltenham. I found none but I learnt that a number of men I had known had migrated with their families to the Isle of Man. I advertised in an Isle of Man paper for skilled engineers willing to work on the island and this brought a response from which it seemed worthwhile to open a factory there. But then officialdom reared its ugly head, for we found that the shipment of machine tools by sea was prohibited. It would be no exaggeration to say that we owe our survival as a nation to our native capacity to outmanoeuvre bureaucrats. The happy outcome illustrates this truth. It was found there was no order prohibiting the export of machines in parts so the machines were dismantled and sent over to the island in pieces. Free from the unwelcome disturbance either by bombs or bureaucrats, this factory did excellent work throughout the war.

My connection with the Isle of Man has been a long one. My brother Robert had started a business there after the First World War. My brother's business was dependent on the tourist trade and completely vanished at the outbreak of war. He found himself with extensive premises unoccupied. It was here the business was set up. The most important man to engage was a machine shop manager and I was indeed fortunate to find a brilliant man for this job, a Mr. Hodson from the National Gas Engineering Company. He was equipped in every department and throughout the war operated the company in an exemplary manner.

He knew exactly what could be produced from every machine, did all the tooling and knew the time needed to execute every job. He never failed to deliver on time and the quality could not be faulted.

On my first visit to America a company, Dowty Equipment Corporation, was registered. Mr. Niedelman, president of Aviquipo, had an assistant a Mr. Ullman and when Mr. Niedelman made it known that he was leaving for a prolonged stay in South America, it was agreed that Mr. Ullman should take care of our U. S. interests. A small factory was set up on Long Island financed by Dowty Group and myself, with an investment of $110 thousand and $90 thousand respectively. There was no other financial interest. This company expanded rapidly through

U.S. government contracts, with the government supplying all machine tools on loan and making advance payments against contracts placed. These were, in general, for Dowty products developed in the U.K.

For the year ended December 1942 sales were $4 million and for the following year, $10 million. The profit shown was a meagre 1% of sales compared with our Canadian company which showed 20% on sales.

During wartime it was difficult to keep control of these companies. We sent two men from the U.K., one technical and the other commercial, but neither had executive positions. It was through them I was alerted to serious malpractices taking place and caused me to visit the U.S.A. in 1943.

Through American lawyers I employed Peat, Marwick, Mitchell and Company to undertake an investigation. They reported that the officers of the company were grossly overpaid and instanced the production director, a lawyer without business or engineering experience. His salary was $18 thousand, but in addition he drew $15 thousand to cover his legal services. He was also president of Phoenix Machine and Tool Corporation, an organisation set up by the officers of Dowty Equipment Corporation and from this corporation he drew a further $18 thousand a year.

There were other companies similarly involved, such as Ford Brothers of Buffalo who were sent almost all the machinery directed to Dowty Equipment Corporation from the Defence Plant Corporation.

The President was Ullman and he drew a salary of $60 thousand a year with a further $7,500 from our Canadian company, but he was also connected with a man Greer, who drew a large salary for doing consultancy work.

The report complained that prices paid to subcontractors were exorbitant, large sums were paid in advance for work placed without price agreement. The report advocated the elimination of these well paid officers.

The report ended by saying it was Ullman's duty to protect the assets and good name of Dowty but the evidence showed that this was not so, indeed everything seemed to be done which was against the interests of Mr. Dowty and his British company and that both Mr. Dowty and the British company's interests were being seriously jeopardised.

I met the chairman of Dowty Equipment Corporation, who had been appointed without my agreement, a Mr. Henry Breckinridge, a man who had held some high positions in government, and Mr. Ullman on the 12th June in the Waldorf Astoria Hotel. I told them of my dissatisfaction and requested changes of management. They acted as I might have expected guilty people to do by saying what did I think I could do?

I visited Washington to the British Commission and showed them the Peat Marwick report but they refused to help saying it was none of their business.

It was fortunate that the company decided to change its name to 'HUB', the initials of these no good officers. I was glad to be rid of a bunch of American thugs. In all my business experience this is the only country I have found those with no business morals.

With no further assistance from the U.K. this company undoubtedly folded up for I never heard of them further.

In the early days of July 1943, I was exhausted and upset with the problems I found in New York and was badly in need of a rest. I stayed with the Niedelmans at their home in Westport for some seven or eight days. It was quiet and I was able to relax, something I could never have done had I gone to Canada or yet returned to the U.K. There was a boat which I could row into the bay, or at night I could amuse myself trying to catch some of the fire flies that seemed everywhere.

On this visit to the U.S.A. I spent two days at Wright Field, the government experimental station, discussing various technical matters including the caterpillar tracks of our design, then fitted to a Boston aeroplane.

Coming from the U.K. and being concerned with the Gloster Whittle aircraft, the first jet engine aeroplane to fly, I was allowed to see the film of this aeroplane! There is a strange reluctance on the part of Americans to give credit to the developments of others. Throughout this film no credit was given to the U.K. , and it was obvious that many accepted this as an American achievement.

For some time my U.S.A. patent agents were Toulmin and Toulmin of Dayton, Ohio and operated by a Colonel Toulmin. Like other professional services in the United States I found their charges extortionate.

I was talking about this to my Dayton friend, Bob Allen, when he chuckled and told me of a lecture that had been given by Colonel Toulmin some years previously, in which he was unwise enough to say, when a client came into his office he would weigh him up and then decide how much he could charge!

Although North America and ourselves speak the same language we often find words with very different meanings. During wartime we were in need of tools for cutting threads and cabled the New York office for screwing tackle. This created great amusement for in America this description is given to tackle for a very different usage.

Indeed it can all be very confusing, for instance that part of the anatomy they call a fanny we call a bum, whereas to them a bum is a tramp and their definition of a tramp is a low female.

Hawker Aircraft were associated with a Major Wylie in the development of an unusual hydraulic pump called the Live Line pump. This was of the radial piston type with means for reducing the stroke of the pistons to zero at a predetermined pressure, thereby cutting off the pump's delivery.

Hawkers could see great possibilities in this design but they were not satisfied by the progress being made and asked me if I would put my weight behind this work and get it into production quickly. I agreed and my company became a third partner in this project.

I found it difficult working with Major Wylie for he could not separate development and production problems. No sooner did he think of some improvement than he wanted it incorporated in production immediately, even before it had been sufficiently tested, and he had no concern for the inevitable delays in manufacture.

Major Wylie was well known with the Ministry technical men who were senior to me and I think regarded me as something of an upstart. Certainly they seemed to do everything they could to make life difficult. That is why throughout the war years I seldom went to the Ministry.

The Live Line pump was used widely on many different aircraft. One day I was called to a live line pump meeting at the Ministry as a result of complaints made by Major Wylie. I made it clear that the pumps needed for the war effort could only be produced if the flow of modifications was to cease, after all it was not as though the pump were not operating well. For some reason I felt the meeting was against me but as I was walking from the meeting down a corridor I felt an arm around my shoulder. It was that of Air Marshal Sir Owen Jones. I forget what position he then held but he was one of those attending the meeting. He said, 'You were absolutely right, don't take any notice of those silly b's'.

In December 1941 I received a letter from the Ministry of Aircraft Production saying that they were willing to make provision for additional buildings and extensions to existing buildings at Arle Court, and to provide additional plant and equipment for the production of aircraft undercarriages. The total cost excluding machine tools came to a sum of £152,000. The machine tools authorised amounted to £124,000.

I have always got on well with my employees and my relations with trade unions have been good but there were some incidents which could have caused trouble.

When my work force was not more than ten persons and I was fighting for my very existence, everyone had to be a jack of all trades. When one employee became very abusive on being asked to undertake some chore, I sacked him not knowing he was the chairman of our local trade union. There were no repercussions.

One Sunday during wartime, during a tour of our factories by important visitors from the Ministry of Aircraft Production, a well known trouble maker was observed addressing a crowd of work people. As no complaint had been made to me I was angered at this demonstration for it was made to appear there was dissatisfaction in the factory. The following day after sacking this individual, I received a deputation from the shop stewards asking for his reinstatement. This I politely but firmly refused. A short time later the stewards asked to see me again and said they had to make the request for reinstatement but fully agreed with the action I had taken.

I was a trade union member for many years. First with the now defunct Toolmakers and then with the Association of Engineering and Shipbuilding Draughtsmen. I was a paid up member until I started my own business.

On July 12th and 13th 1941 there was a weekend discussion on undercarriage problems in wartime under the auspices of the Royal Aeronautical Society, held at Arle Court, Cheltenham, It is interesting to record the important people that turned up on that occasion, amongst those were Sydney Camm (later Sir Sydney) of Hawkers, W. G. Carter, chief designer of Glosters, Roy Chadwick, chief designer of A. V. Roe, Mr. H. G. Conway of Rubery Owen Messier, Dr. Harold Roxby-Cox (now Lord Kings Norton), Mr. Roy Fedden (later Sir Roy), Mr. H. P. Folland, chief designer of Folland Aircraft, Mr. Arthur Gouge (afterwards Sir Arthur) of Short Brothers, Major Halford of Napiers, Mr. F. G. Miles, head of Phillips and Powis, Lt. Col. the Rt. Hon. J. T. C. Moore Babisson at that time minister of Aircraft Production, Mr. J. D. North of Boulton Paul, Dr. Pugsley of the Royal Aircraft Establishment (later Sir Arthur) and Sir Robert Renwick from the Ministry of Aircraft Production, Oliver Simmonds (later Sir Oliver), my old friend Joe Wright of Dunlops. It was interesting on this occasion to have pilots from the fighter, bomber and trainer commands and their views were most valuable.

During the critical period of the Battle of Britain it was of crucial importance that every single fighter plane we could muster should be put into the air. This meant that every crashed machine must be salvaged, and its components reconditioned and put back into the production line as rapidly as possible. Lord Beaverbrook, Minister of Aircraft Production, sent for me and for the next few days I was to receive several tons of such items for repair. There was no space available for such work at Arle Court, while the urgency ruled out any idea of building a new factory. Some suitable existing building was needed, but none was to be found in Cheltenham. The quickest way of prospecting the neighbourhood was from the air, so our pilot Martin Fountain Barber flew me over the towns of Cirencester and Tewkesbury, where I spotted a large four-storey brick building standing beside the railway station at Ashchurch. This had originally been built by the Midland Railway Company as a forage store for railway draught horses and was then partly occupied by Birds Custard Powder who had been bombed out of Birmingham. Within one hour I was inside the building and found it had a considerable floor area not utilised. The government speedily agreed to its lease and so by this unorthodox fashion there began an association with Ashchurch which was to lead to remarkable developments after the War, and to the employment of many thousands of people.

In this new repair factory the salvage and repair of Dowty components were quickly put in hand and a great deal of material and equipment was saved.

When we took over the Ashchurch factory we acquired about 40 girls, whose experience until then had been making custard powder. In order to give them some preliminary training we persuaded the engineering department of Cheltenham Technical College to take them for a two weeks' course, but no sooner had they arrived than we had a call from head of department saying these girls were wrecking the place and quite out of control, and we must take them away. We only had two educational films to show them and at the end of the fortnight we could say that we had the only 40 girls in the world who could pack custard powder and read a micrometer. Soon machines were installed at Ashchurch and the girls given practical training. All floors were connected with an internal staircase on one side of the building and a fire escape. There were numerous outside air-raid shelters and close by was a U.S. depot with 4,000 G.I.s. The clever use of all these facilities coupled with the blackout provided situations which no personnel department could control, for a spirit of adventure was abroad!

Our Tewkesbury interests during the war were well served by a Mr. Gaze, one-time Mayor, and proprietor of the Bell Hotel. He was our PR man in the district. I passed his hotel on the way to our Tewkesbury factories and would call in there for a meal without the improper use of petrol. A most satisfactory arrangement!

In recalling wartime happenings I remember the wife of an employee who wrote to me saying that as her husband was always working on the night shift she was very lonely and could I do anything about it!

In May 1942 Colonel Llewellyn, the Minister of Aircraft Production, set up a joint investigation on the best ways of securing additional undercarriage manufacturing capacity and this was carried out by Sir George Nelson (later the First Lord Nelson of Stafford) and myself.

At that time my company was producing the landing gears and hydraulic components on a fantastic number of aircraft . To give some idea of this, the Lancaster programme included 96 complex products and orders against each of these items ran into several thousands.

With this big programme entrusted to my company we were told that the Messier company, who were only making undercarriages for the Handley Page 'Halifax', could not meet their commitments on this one bomber aircraft. The Minister sent Sir Robert Renwick to ask me to [produce] one half of the bomber undercarriages for the Halifax aircraft. It was then that the Ministry of Aircraft Production thought we should take over a factory at Corsham.

On 23rd June a meeting took place at the Ministry of Aircraft Production with Sir Archibald Forbes in the chair, and as a result a meeting took place at the Bristol Aeroplane Company at which were present Mr. Verdon Smith (later Sir Reginald) and Mr. Rowbotham of Bristols, the director of machine tools and myself, and it was then decided to take over a large part of the underground factory of Corsham in Wiltshire to produce 125 sets of Lancaster undercarriages and 125 sets of Halifax undercarriages per month. The Bristol Aeroplane Company agreed to be responsible for the ordering of the machine tools and materials and a liaison office was to be set up to give Bristols all the assistance

George standing next to a Lancaster undercarriage test rig

they needed from us on technical matters, the procurement of machines, materials, drawings, etc.

The lack of co-ordination with the Ministry of Aircraft Production was amazing, for soon five wagon loads of Grindley automatics were delivered to Cheltenham Station, further wagons were addressed to Corsham Quarries, Cheltenham, there was complete chaos.

The factory area at Corsham was no less than 2. 25 million sq. ft. of which 400,000 sq. ft. was allocated for undercarriage production but no provision for the installation of machines had been made so they were temporarily set down in the wide roadways.

In July that year a director of undercarriage production, a Mr. Gwynne, a hotel proprietor, was appointed. This man visited me on one occasion but I was not impressed for he was no engineer and without knowledge of the problems. Without reference to me this man called a meeting with the Bristol Aeroplane Company and cancelled the entire Corsham project. As a result I wrote a letter to the then minister of Aircraft Production making the strongest possible complaint and referring to the written arrangements made between the Ministry and my company that we should always be consulted on matters affecting our contracts. I said I had no confidence in his director of undercarriage production. I received no reply to that letter and I never saw the director of undercarriage production again.

I called at the Ministry and said that since they had cancelled the Corsham scheme it was their responsibility to find alternative capacity. The Ministry sent a high powered delegation to Messrs. Levers of Port Sunlight and to Metropolitan Vickers requesting them to take over this work. They did so but only under protest. Twelve months later Metropolitan Vickers had never provided a programme or given any deliveries and the schedule on Levers was never realised. I appreciated how difficult it was for these companies to come into this programme at such short notice.

Early in 1943 I found myself saddled with a government enquiry into the operation of my company. Mr. J. K. Weir (now Lord Weir) and Mr. Jamieson, Production Director of Messrs. G. and J. Weir visited me on the 1st of February, on instructions from Sir Stafford Cripps. They came to enquire into the quality of our management and ability to meet the heavy programme placed on us.

The committee asked many searching questions, interviewed members of my staff and work people, and visited many of our sub-contractors. All these investigations took many days.

On the day that Mr. Weir left he asked me to dinner and on leaving he said that if I would go to my office I would find a book on my desk of standing orders with a marked page. That, he said, would tell me who had been responsible for the investigation. This was the man Evans Hemming employed by the firm of consultants I was using.

Later that month I received a confidential report from Mr. Jamieson saying the brief they had been given dealing with the conditions in my company had been most misleading and many valuable hours had been lost in disproving the false statements contained in the brief.

The report said it was pleasing to record that the investigator was informed generally by people in the factory, that Mr. Dowty could be seen at any time should grievances or misunderstandings arise and that the relationship between employees end the management was good.

It was pointed out that 90% of any problems were entirely due to the late delivery of raw materials which was not our concern.

Mr. Jamieson finished his report by saying that bearing in mind the position of our company a few years ago, we had accomplished a magnificent task and were to be congratulated on our efforts. He was definitely of the opinion that we would meet the requirements of all the aircraft producers during the coming months.

I heard nothing further of this investigation.

Cedric Howarth was programme controller on undercarriages. Every few days there was a new programme. The Weir Commission referred to the impossibility of keeping up with the weekly changes of programme. More papers and forms flooded in, all cancelling the previous issues. But we had our own people working closely with the airframe manufacturers and knew exactly what they wanted weeks before the programmes came through the government departments. This initiative on our part was appreciated by the aircraft industry but disliked by the civil servants who wanted to appear all important.

Throughout the war we kept the aircraft industry supplied with all their landing gears and hydraulic equipment which in total represented greater effort than any one aircraft manufacturer. Every item manufactured was of our design. No aircraft programme was ever delayed for lack of our equipment and no single aircraft was ever unserviceable for lack of spare parts, a record that perhaps was unequalled.

This could not have been done without using our initiative, but for this we were often subject to criticism, and this was reflected in that no official recognition was ever made for this mighty contribution to the war effort.

Sir Stafford Cripps was a frequent visitor during the war. I felt he used some of these visits as an excuse for he lived on the Cotswolds, not far away.

He generally stayed to lunch which we found somewhat of a problem as he was a strict vegetarian. He would often use the microphone in my office to address our employees over the intercom. As an avowed Socialist he was always preaching the equality of sacrifice but meanwhile the black market flourished.

In October 1943 I received an all day visit from Air Chief Marshall, Sir Wilfred Freeman – chief executive of the Ministry of Aircraft Production. I felt that his visit was in some way to acknowledge the good report made by the Weir Commission.

At the end of May 1943 I travelled on a U.S. Air Force Command aircraft, a Douglas C. 54 Skymaster, to the United States. This trip took us via Reykjavik, Iceland and our total flying time to New York was 18.5 hours.

Leaving England one day and arriving in America the next, I was able to make direct comparison between wartime conditions in these two countries. I was struck by the almost complete lack of restrictions. The shops were full of commodities, no clothes were rationed with the exception of shoes but if one went to a shoe shop and had no coupons these could be purchased, so even this rationing was completely defeated. There was no food shortage. I could have as many courses as I wished with as big a variety of food as in normal times. There was supposed to be two meatless days a week but one was never refused meat. The only beverage rationed was coffee, but even then you could have as many cups of coffee as you wished and sugar was in abundance.

Motoring was supposed to be restricted but when I went to the beaches at weekends these were packed with thousands of cars.

There were no blackout restrictions except on the coast and the New York skyscrapers.

The American way of life seemed quite unaltered but one thing amused me and that was the restriction on the number of flavours permitted for ice cream. These had been reduced to twenty-nine.

Unlike conditions in the U.K. there seemed little control of costs or wages and there were splendid opportunities for making large wartime profits.

In the factories I noted the small percentage of women at work. Able-bodied men were operating small capstan lathes and doing menial jobs which in the U.K. were done by young girls.

Night life in America was that of pre-war with the night clubs as numerous and running till the late hours of the morning. The first of those wonderful musicals 'Oklahoma' was breaking all theatre records.

On my visit to Canada I saw many things I found peculiar. In Montreal it was not possible to get a drink in a bar when standing up, drinks were only served to those sitting. One day I was taken to see the factory of Bob Noorduyn, the designer with whom I worked at British Aerial Transport during the First World War. I passed four complete race tracks, each with its grandstand, paddocks and stables, all on one plot of ground. It seemed incredible that anybody could have been guilty by quadrupling what was obviously a very large capital expenditure. The reason for this I was told was legislation which made it illegal to have race meetings on any single track in Canada for more than two weeks in each year. To overcome this law, they built four race tracks, so were able to have eight weeks of racing.

Having spent eight weeks in North America I left for England on July 31st. We stayed in Bottwood for twenty-four hours as we were told it was a bank holiday in the U.K. with no air controllers!

In 1943 Sir Roy Fedden, having completed another of his missions to the United States, issued a report covering American landing gears and power-operated services for aircraft. Sir Roy was pro-American and could see nothing but good in everything they did. His report covered comments on items about which he had no expert knowledge and condemned, for instance, the use of hydraulics for the operation of retractable undercarriages and advocated electrical means.

I found this unfortunate and issued a printed twelve-page report showing these proposals to be quite wrong. Had I not done so I felt someone in government circles might believe Fedden's comments to be true and that, as a country, we were slipping back. Sir Sydney Camm said at our company's Jubilee Luncheon, 'We had to decide at one time whether to operate retractable undercarriages electrically or hydraulically. George Dowty said to me, "it seems odd to move something which moves at less than one revolution per minute by something that goes thousands of revolutions a minute"– so we decided on hydraulics and twenty-five years later I know that decision was right'.

There were those who were concerned at the lead the U.S.A. were gaining over us in the development of civil aeroplanes and l was asked to serve on a committee which included Sir Peter Masefield, to consider our future position and to make recommendations to the government.

Throughout the war years we operated a school to train British and Allied persons in the operation and maintenance of hydraulic equipment. Over 8,000 students attended this school to gain information which could not have been obtained elsewhere.

In 1944 I took out patents for what has been a real money spinner – the bonded seal – a metal ring with a moulded rubber ring of special shape. This seal has been produced in hundreds of millions and made under licence in Germany, France and the U.S.A. In this year I also patented the hydraulic ram with internal claw locks – designs that are still in use.

In 1945, my group set up a small plant for the manufacture of rubber seals in St. Mark's Parish Hall. The equipment was a small press operated by a 'Lancaster' bomb door jack and an 'Anson' hand pump. There was one mixing mill with rollers 12" long and 7" in diameter, a table a pair of scales and dustbins to hold the ingredients to be mixed. The first bonded seals were produced at St. Marks and later transferred to a site in Bath Road where five people were employed. By 1947 production of bonded seals reached 8,000 a week and production was then moved to Ashchurch and has been there ever since. No-one could foresee this manufacture of seals was to expand into many factories employing thousands of employees.

In addition to the worries of operating an essential wartime business there seemed little rest at night. To give me further help my twin brother, who had businesses in Blackpool, came to Cheltenham to assist me. We were identical twins so that his coming gave rise to many amusing incidents for he was always being mistaken for me.

On January 22nd 1945 a great tragedy occurred. We were both travelling to London by car with Sir Roy Fedden, Dick Spires, our chief service engineer, and my chauffeur Oliver, when some miles before Oxford we were struck sideways by a heavy army lorry out of control. It hit our car sideways pushing it off the road and down an embankment. My twin brother sitting on the outside took the full force of the impact.

He suffered severe injuries. I was sitting between my brother and Sir Roy and came out better than the other passengers for I was able to walk a mile or so to a nearby cottage and telephone for help and speak to my brother's wife. I then returned to the scene of the accident and travelled in the ambulance with my brother and the other passengers to Oxford Infirmary.

My brother suffered with a heart condition for which he took pills but the hospital staff, in spite of my protestations, took the pills away from him which made me angry. Although I was terribly shaken I was able to travel home but had to take to my bed. The injuries I received did not make themselves immediately apparent My brother asked for me and I travelled to Oxford but I was only in time to see his body wheeled to the mortuary.

Three days later he was buried in Pershore in a grave next to our father and mother. First there was a service in Pershore Abbey, where we had both been christened. After the service my brothers tried to restrain me from going to the interment but I insisted on going for I felt compelled to do so for one who had been so close to me all my life. The loss of my identical twin was a most terrible shock.

Some time later there was an inquest at Gloucester, an ordeal that I would have given much to have missed. We had grown up together, we thought alike, we spoke alike and had the same characteristics and mannerisms. Our lives were uncannily similar. His loss affected me greatly and after thirty years his memory is as strong as though he was still alive.

On Trinity Sunday in June 1946, a special service was held in Pershore Abbey where altar rails I had caused to be erected in his memory were dedicated by the Archdeacon of Worcester.

I regret that some years later a vicar, who wanted to give the chancel a new look, had these rails removed. Some were placed in a side chapel but I found the remainder stored in a room off the belfry.

My twin brother was an accomplished musician and had been organist at St. Mary's and St. John's in Worcester. It was fitting that this service included an organ recital on an instrument which had been completely renovated in recent years in memory of his father and mother.

As a result of this accident Sir Roy Fedden was also confined to hospital and two weeks later I developed phlebitis and pleurisy when staying at the Savoy Hotel in London and was well looked after by a Dr. Harry Mason.

The burdens placed on me during wartime, the air raids, sleepless nights, domestic problems and the loss of my twin brother were too much. One night about 10.00 p.m. after chairing an all day progress meeting I suffered a complete mental collapse. It seemed as though my brain had caught fire, I did not know where to place myself, I was demented. I placed my head under a cold water tap, I smoked incessantly, I was desperate.

About this time an American woman doctor – a sister of Mrs. Hilda Niedelman of New York – was in England attached to the U. S. Forces. It was through her influence I went to a Dr. Henry Wilson in Harley Street but while I was there I was so ill that I went about clutching a piece of paper with my name and address in case I collapsed, for that is exactly how I felt and I can understand those who take the easy way out. Eventually I was sent to a Dr. Lincoln Williams at Harrow who ran a nursing home dealing with cases such as mine. It was there that I underwent sleep treatment and I was under the influence of these drugs when the war ended. When I came around I was terribly depressed and low spirited. After a week or two getting fit enough to travel I returned to Cheltenham but was back in the nursing home three days later.

I tried to get away from business by visits to Torquay and Tenby but they gave me little relief. Later I went to stay at the 'Rising Sun', an inn at the top of Cleeve Hill. But I could never stay away for long.

The war being over there were the major problems of business reorganisation to meet completely new conditions.

Our wartime sales reached £10 million per annum but the government imposed 100% taxation on profits in excess of the meagre profits made in the pre-war years. This confiscatory tax on profits prevented the payment of any commission to me, but this arrangement should have been reviewed at the end of hostilities. However, at that time I was suffering from a severe nervous breakdown. This resulted in my private affairs being neglected and those who might have helped did not do so.

I had another good friend at the Ministry in those days in Major C.J. Stewart, a man of great experience who had been superintendent of the Royal Aircraft Establishment and during the war was controller of production at Ministry of Aircraft Production. At the end of the war he joined our company to give help in the development of our aircraft fuel systems. It was a great loss when in 1954 he died suddenly while on holiday in Scotland but he had lived to see the fuel system grow into a separate entity in 1953.

The desperate plight of the Germans after the war was reflected in letters from Herr Brauer, our old agent, who amongst other things asked me to send him shoe laces. In England there were shortages too and I set up a factory to make metal suitcases, plastic braces, under the name of 'Arlex'. These products were not in our line of business and the company was ultimately sold but is still in operation thirty years later.

Post War

BEFORE THE WAR ended the post war future of my company had been a matter of very anxious concern. As our profits had been wholly taxed throughout the War there was the prospect of changing over to peace time trading without financial reserves. We had been almost exclusively in the aircraft industry and it seemed all too probable that this industry would decline as it did after the First World War. I well remember that then many aircraft companies went into abortive schemes on which they lost heavily, but surely, I thought, we had accumulated under the stress of war, a kind of wealth which could not be taken away – a wealth of knowledge. It was then that we looked to see how we could profit from the techniques we had developed. We thought of ways in which this knowledge could be applied to other industries. Unlike companies that had been soundly established before the war we came out of the war without capital for, as I have said, we had not been allowed to make profits during wartime. But I have always maintained that good ideas are more important than money and this was to be shown once again when I record our post-war developments.

The most outstanding was our entry into the mining industry followed by the development of synthetic rubber seals for the hydraulics industry and the application of hydraulics to many industries.

The end of hostilities brought about wholesale cancellation of contracts and we found ourselves with many millions of pounds invested in complete undercarriages, components, parts and raw materials. We took over large warehouses in which all these stocks were housed and it was due to good housekeeping that in later years we were able to utilise much of this material.

Our Chairman, Mr. A. W. Martyn, had suffered ill health and died in January 1947. From his conversation with me on his last visit it was clear he did not expect to see me again. I owed him much for his support at a time when others would not help although this had not prevented my small company being forced to go public far too early.

In 1947 Roy Chadwick spent a weekend with me in Cheltenham with a model of his latest design, the 'Vulcan' bomber. He was not a happy man at that time and confided in me that he thought of joining Boulton Paul Aircraft. Unfortunately, he met a tragic death the following week when flying in an aeroplane where the ailerons had been reversed. With him on that flight was Stuart Davis who has twice been in my employ. Davis saved himself by the foresight of bracing himself between the walls of the toilet.

After Roy Chadwick's death I was embarrassed to find he had left practically all his money invested in my company which, I suppose, was a compliment. However, this fortunately turned out well for his family.

I am indebted to his daughter, Rosemary, for providing me with interesting facts about Roy's life and how, in his early teens, he was a fanatical builder of gliders and elastic driven model aeroplanes. So many of us came up the same way by receiving inspiration from aviation in the early days.

In a broadcast he gave in 1942, he said to young engineers that if they took up aeroplane design they could find it the most fascinating, absorbing, exciting and heart-breaking business and they would require no other hobby to occupy their spare time for they wouldn't have any. That is the way these great men built up their businesses but I cannot see the same enthusiasm with young men today.

Those who worked at Hamble in the early twenties knew Roy Dobson (later Sir Roy) affectionately as 'Dobbie'. He then ran the Manchester factory of A. V. Roe but was later in charge of that company's operations.

Some time elapsed before the 'Vulcan' designs were sufficiently advanced for orders to be placed. One day I was asked by Dobbie to meet him in the Hawker Siddeley office in St. James's Square, London, and was shown the massive documents submitted by one of our competitors and the relatively scanty submission made by my company. I felt somewhat embarrassed. After some discussion Dobbie said 'You have done good work for us over many years but we know nothing of these other people. I am sure you will do a good job so I propose over-ruling those in my organisation who would make another choice. You will get the business. ' He had confidence in me and a respect for my work.

Over the years Dobbie and I were great friends and it was he who proposed that I should become the first equipment supplier to be elected president of the Society of British Aircraft Constructors, a position previously held by the many aircraft and aero engine constructors.

As the Canadian company was wholly engaged on aircraft products its activities declined with the ending of the War. It was then that they employed a firm of market research consultants to look into new products, one of which was an automobile hydraulic jack. The reports were wholly favourable so that designs were developed and productionised. In due course the product was marketed but immediately met severe competition from the U.S.A. , a lesson that a market research report is good only at the time it is written. A few weeks later its findings can be invalid.

Until this time we had started all our companies from scratch but in 1948 we took over a small engineering company, New Mendip Engineering of Atworth in Wiltshire, owned by the brothers George and Arthur Thacker. They had a poor machine shop but with great ingenuity produced first-class work. We were able to take advantage of that company's tax losses so acquired the business for a song.

For three years I struggled against ill health but in March 1948, I realised I could do so no longer and, fortunately, I took a long rest. I chose the island of Nassau for I knew Fred Sigrist late of the Hawker Company, who had emigrated there before the war. For three weeks I never left my hotel suite. After that rest I went deep-sea fishing with Arther Hogg, chief designer of the De Havilland Aircraft Company, who I found was also staying in the hotel. In two months I felt better. Towards the end of my stay in Nassau I met my wife-to-be. However, I was not completely cured and as late as 1954, eleven years after my first breakdown, I had a recurrence of my trouble and had to take another break from work in Madeira. Because of my continued ill-health I bought an extensive property, Grayswood Hill, on the outskirts of Haslemere where I spent long weekends. It was a splendid residence with most beautiful gardens, a sheer delight. On the 16th July 1949, I was married there with my wife's relations and friends coming from Canada. Some years later, with a young family, I found travelling between the West of England and Surrey a chore and reluctantly sold the property to a member of the Pilkington family.

For some years we had been investigating the possible new applications of hydraulics to other industries. Looking at coal mining this seemed a fruitful prospect since the mining of coal was being carried out by the time-honoured methods of pick-axe and shovel with wooden pit props which had to be cut to length and sledge-hammered into place. This was a crude method of extracting coal.

Towards the end of the war our chief service engineer, Martin Fountain Barber, who had friends in the mining industry, came forward with a design for a mine tub retarder. He followed this with a suggestion for a hydraulic pit prop. This was not a practical design. It had, for instance, a long slot through which coal dust could enter. I produced another solution, the prop with the well-known offset boss to accommodate a lever large enough within the telescopic tube to operate a hand pump, and this had integral means to effect its release. I took out a patent for this in 1948 and for the first hydraulic chock in 1949.

Since the earliest days of mining methods of coal extraction had not changed and it was apparent that a hydraulic pit prop could be set in position and removed much more quickly than wooden props. Furthermore, the hydraulic prop would be provided with a valve so that should a prop become overloaded it could close slowly to allow the load to be shored by the adjacent props. At first it seemed that the high cost of a hydraulic prop could rule out its adoption but whereas wooden props were expendable the hydraulic prop could be recovered and used repeatedly. In fact its life was many years whereas the wooden prop could at the most be set only twice, and the cost of timber and its transportation down the mines was expensive.

There were great difficulties to overcome selling this product to the National Coal Board. Hitherto the cost of timber props was written off as soon as they were purchased, for their life was only a matter of days. There was no method in the accounting procedures for treating hydraulic props differently. Mine managers were reluctant to purchase hydraulic props and write them off the day they were purchased! Although hydraulic props cost fifty times as much as wooden ones they lasted six hundred times as long, but no account could be taken of this substantial advantage as it would upset their accounting system!

Perhaps the most important advantage of the hydraulic pit prop was its safety factor, since its use could prevent the fall of the roof, the cause of many fatal accidents. The hydraulic pit prop became the most outstanding contribution to mining safety and efficiency since the earliest days of coal mining. However, it took many years to convince the mines of these advantages.

Although the hydraulic pit prop made life easier for the miners, yet they demanded extra payment for handling this new equipment.

The fruits of success did not come quickly for, faced with the prospect of nationalisation, colliery owners were unwilling to spend capital on new developments. Several years elapsed before the first enquiry for 3,000 props came from the East Midland division of the N.C.B. From then on the sale of props accelerated such that a complete manufacturing plant for mass production was laid down in 1956.

This prop was an outstanding development that has since been copied by many others. The hydraulic support for mine roofs is now used throughout the world.

Many manufacturing licences were granted to overseas companies and one must pay tribute to the late Robert Lancaster of the N.C.B. Welsh division, who encouraged these developments. He frequently visited me and it was due to his suggestion that we produced the first advancing roof support, but that is another story.

In 1948 I set up Dowty Mining Equipment and another company that was to become Dowty Hydraulic Units. In that year we suffered another loss when our works director, Mr. V. O. Levick, who had worked with me in the early days in Hamble, was drowned while he was on holiday.

In 1949 the Canadian company moved from Montreal to Ajax on the outskirts of Toronto and at that time Mr. R. F. Hunt went to Canada to take charge of our Canadian operations.

Canada, I found a rough country in many ways. Its hotels primitive by European standards. Every bedroom has a T.V. of poor quality from which outpoured an endless stream of commercials. Staying at the Royal York Hotel one Sunday I noticed yet another convention for which North America is noted. Enquiring as to which this was I was given the one word 'Alcoholics'.

Toronto I found a rather drab dull city with commonplace dwellings stretching enormous distances from the city centre. It was a city that can be aptly described as 'miles and miles of undulating bugger all'.

I find Toronto a depressing place where most of the populace seem to eat their meals sitting on stools at a drug store counter, whilst the more opulent of the community belong to and eat at exclusive but stodgy clubs.

There was one innovation I found of interest, the large dome standing high over other buildings. It changed its light according to the weather expected and if the light moved up the dome it denoted a rising temperature and if downwards the reverse. I found this method of weather prediction unique.

In business there has grown up a big racket becoming enormous by making everyone believe that the risks that are run in every way are far greater than they really are. This is shown by the profits made by insurance companies. In Toronto, for instance, all the skyscrapers are owned either by banks or insurance companies.

Every company has a proliferation of guards, commissionaires doing duty on doors, Securicor men and the like, all making a living in the belief that the dangers they are there to prevent are greater than they really are.

One risk is in the wages taken by companies from the banks each week and is something that could be avoided if wages were paid by cheque.

In the early fifties I had considerable dealings with that important Cleveland, U.S.A., company, Thompson Products (now Thompson Ramo Woolridge) or TRW as they are called for short.

In those days the head of the company was Fred Crawford, a great character. On visiting England on one occasion, I took him to the Farnborough Air Show and it was then that he described his company as being the nut and bolt people of America. In point of fact they were the largest manufacturers of automobile engine valves which they turned out by hundreds of thousands daily.

When they went into more exotic activities such as those in which we were interested, they were not so successful but we had a happy association over many years.

The man in charge of their engineering activities was Arch Colwell, a charming man with whom I had the most cordial and happy relationship. He also was a frequent visitor to Europe. In fact I met Arch Colwell and Fred Crawford in many parts of the world and only a few years ago at Lyford Cay in the Bahamas when we were on holiday.

The manner in which we became involved in jet engine fuel systems goes back to 1943 when the De Havilland Engine Company used our Live Line hydraulic pump as a fuel pump on their first jet engine, the 'DH 1'.

This was designed to pump oil but was now required to pump a non-lubricating fluid. It was not until 1945 that after considerable development it passed its 150 hours endurance test.

At this time I held a meeting with the late Major Halford and Dr.

Moult of the D. H. Engine Company when it was agreed we should build up a facility to produce complete fuel systems. From then developments were rapid and we were soon producing systems for all D. H. engines used on such aircraft as the Gloster 'Javelin', the Handley Page 'Victor' and Fairey 'Gannet'.

We were fortunate in finding men and equipment capable of producing work to a consistently high standard. Mr. Fred Carey who had been in charge of this work was acquainted with the Thatcher Brothers of the New Mendip Engineering Company in Wiltshire. They were brilliant production engineers and we bought their company in 1948.

Until now all this work had been carried out as a department of Dowty Equipment.

In 1950 we set up two other companies, Dowty Seals and Dowty Fuel Systems, both of which were to grow into important businesses. The following year we acquired our first outside company of any size, Coventry Precision, one of our good wartime subcontractors. It owned an extensive freehold site, employing 700 people with first-class machine shops and a row of houses. This was a factory once owned by Sir Harry Harley but it had since passed into the ownership of the Tecalemit company. Sir Harry was still interested in that business but whereas he and I got on extremely well I sensed this was not the case with Tecalemit.

He approached me to suggest buying the business at the extraordinarily low price of £100,000. I accepted this offer with alacrity but I could never understand how the Tecalemit company sold this factory for so paltry a sum. I was greatly indebted to Sir Harry who unfortunately died a few months later on a voyage to Australia. It was then that Sir Harry's son, Stanley (now Sir Stanley) joined the board of Coventry Precision.

It was most fortunate that the man in charge of this factory was one of those treasures brought up the hard way by Sir Harry. He knew every machine and operator and was the most experienced of machine shop managers. While he remained in charge the business prospered but unfortunately he died some years after we bought the business and from that day its fortunes declined. There seemed no replacement for such a man.

Short of technical men at the end of hostilities in desperation I advertised in the Canadian papers and received hundreds of applications from those wanting to return. A team went to Canada to interview these applicants. The main reasons for so many wishing to return was a desire to have their children educated in England, the high cost of hospitalisation in Canada and the womenfolk, who had left England, wanted to return home. It was at this juncture we took over the mansion of Brockhampton Park, Andoversford to house twenty families from Canada for no accommodation was available in our district at that time. This turned out well.

Mr. A. W. Mills came to us from Rayrolles of Newcastle and was a very popular manager of our Ashchurch factory. He acquired a nice revolving chair of which he was extremely proud. If a meeting had gone well he pushed his chair back, spun round with his legs raised and said, 'That's it boys. ' Everyone was most impressed. But his last performance was the one everyone enjoyed most, the chair spindle broke and he shot under his desk. He never repeated the trick.

In 1955 I took over a small company, Davis Wynn and Andrews, a business concern with the instrumentation of nuclear power stations. They had some financial problems but had a young enthusiastic team of men from the aircraft industry and as such were men with good training. Their leader was a Mr. T.D.H. Andrews, a very competent and dynamic engineer. A new company was named Dowty Nucleonics and moved into Brockhampton Park. An important contract undertaken was the instrumentation for the Berkeley Nuclear Power Station. After a year or so it appeared that the company was largely concerned in making submissions for nuclear stations which were doomed never to be ordered.

At that time I was chairman of the C.B.I. Nuclear Committee and in that position saw the gradual reduction in the number of nuclear power stations to be ordered. Starting at twenty-four it eventually finished up at six and it was apparent that too many people were chasing too few contracts.

At the end of 1959 I asked the staff of Dowty Nucleonics if they were wedded to their particular work or if they would like to put their activities to the development of products which could be of interest to our various companies. They were delighted with this proposal and that is how we came to change the name of the company to Dowty Technical Developments.

In 1952 I was happy that my wife gave birth to a daughter, Virginia, but in the same year I was unfortunate to suffer another great loss in the death of our production director, Barney Duncan, another of my associates at Hamble in the early twenties. He came into my office one afternoon looking desperately ill complaining of pains in his chest. I telephoned his doctor expressing my concern but he laughed and said it was indigestion. Twelve hours later Duncan was dead!

That year I took over the chairmanship of the North Gloucestershire Disablement Committee, a position I held for eleven years. During that time we worked exceedingly hard to place handicapped people in employment. In my own factories we did everything we could to find jobs for these people even to the extent of employing blind inspectors using Braille micrometers. Due to much hard work we were able to claim we had the lowest figure for disabled unemployed per head of population of any other area in the country.

1952 saw the coming of age of our business and a book published at that time gives the personal recollections of three of my early employees. John Dexter said that in the early days at Grosvenor Place South work never seemed to stop and even when they moved from that location to Arle Court, each man packed the machine he was working on, travelled with it, unpacked it, set it up and started work on it the minute he could. Such was the enthusiasm of those days.

Joe Bowstead had a small lathe in the cellar under his house and earned a bit of spare cash with it in the evenings. It was Joe who helped Fred Kilner, another of my early employees, to make the first exhibition panel for the 1933 R.A.F. Display at Hendon. He recalls that they worked overtime at night and were often half asleep but they stuck to it so that the exhibit was ready on time. Another great character was Arthur Walton. He said, 'The packing cases I've nailed up for Mr. Dowty would fill the Albert Hall, large ones and small ones containing anything from a tail wheel to my own lunch! But with all the hammering I've never hit my thumb once. I remember he said the Bristol Company urgently wanted a tail wheel. I put it on the train at eleven o'clock one Saturday morning. By one o'clock Mr. Dowty asked if it had gone because Bristol's telephoned about it. On Monday and Tuesday as Bristol's kept on telephoning I showed Mr. Dowty the delivery sheet signed by Bristol's and the return case. "Well open it. " he said. And there was the wheel. You would hardly believe it but when I sent if off again it came back a few days later still unopened. But this time I sent it off again without the case and then it did not come back.'

When Walton first called to ask for work he was 60 years of age but I gave him a job and at 81 he was still working and did not want to retire.

I had joined the Royal Aeronautical Society as a student member in 1918 and in 1952 was elected their president, the only one to rise from the ranks of student other than my old friend Lord Kings Norton.

I gave much time to the affairs of the society. For several years I was chairman of their publications committee. Their journal had a funereal appearance but I was able to get this completely modernised and as a result, obtained a considerable increase of revenue from advertisements.

Their financial reporting was poor and the treasurer at that time appeared to be interested in concealing the Society's affairs and it was a hard fight to get this altered.

In 1967 I was awarded honorary fellowship of the society, a distinction granted to but thirty out of a membership of over 12,000.

The major society lecture each year is the Wilbur and Orville Wright Memorial Lecture given alternatively by U.K. and U.S.A. lecturers.

As the society has branches throughout the Commonwealth it was on my suggestion that there should be a British Commonwealth lecture and this was inaugurated in 1945 and is given alternately by lecturers from the U.K. and the Commonwealth.

For twenty-six years the society's secretary had been a great character, Captain J. Lawrence Pritchard. After his death Sir Peter Masefield wrote in the July 1968 issue of the *Aeronautical Journal* that Pritchard made no secret of the fact that of all presidents he had served I took pride of place – a handsome compliment.

The following year Dowty Group Limited was formed. The name Dowty Equipment Limited had become too much associated with the aircraft industry and our interests were now widely spread.

I was always anxious to help the British aircraft industry when I could. An opportunity occurred about this time. When Mr. Christopher Clarkson left his post as Civil Air Attache in Washington I felt he was the man best to represent the British aircraft industry in the United States, but at that time the industry would or could not support him. I put Clarkson on the payroll of our American company and he remained with us for two years until he was successful in obtaining the U.S. contract for Viscounts and then Vickers took him on their full time staff.

At eleven o'clock in the morning of 13th April, 1953 I was delighted to hear of the arrival of my son, born in an Oxford nursing home. He was named George after myself and Edward after my twin brother. He did well at his studies and sports, was educated at Dean Close Junior

School where he was Captain of Cricket, and then went to Marlborough where he was later joined by his sister when that college became co-educational.

Both my children are accomplished skiers having been to winter sports in Switzerland since they were very young.

Having taken an interest in my business and my work all his life, my son opted to take up engineering as a career and my daughter went to Oxford to read Japanese. After two years she gave up this course to be married.

In 1954 I was elected the first Honorary Fellow of the Canadian Aeronautical Institute. My old friend Harry Folland died in September 1954, at the comparatively early age of 65. He started work in the Automobile Industry and then joined the Royal Aircraft Establishment, Farnborough, coming to the Gloster Aircraft Company as chief designer after the First World War. Harry Folland always gave me his full support and used my products on the aircraft he designed. I shall always be grateful to him for the help he gave me in the early days of my business. When Hawker Siddeley bought the Gloster Company he founded Folland Aircraft in Hamble and produced that very successful aeroplane, the Folland 'Gnat'.

In 1955 I was invited by Mr. Walter Monckton to join the board of Remploy, on which I served for four years. It was not altogether a happy experience for there was no desire for the company to become a business like operation. They could not get first class managers for they would not pay the rate for the job. I decided to take over factories to show what could be done. We sponsored these by providing material, work, machines and management. I believe these were the only factories to show a profit! I wanted sponsorship of more factories by Industry which meant the employment of a top-class executive to visit the chairmen of the larger companies to enlist their help. But this was thought not to be a good idea and I was left with a feeling that so long as Remploy continued to make heavy losses it was regarded as doing a good job. I could not accept this and like others before me, I resigned from the board. Of the two factories we sponsored, the one at Bristol is still running after nearly twenty years. The other, for the repair of mining equipment, was closed when the N.C.B., in spite of our protestations, placed this work in their own workshops.

It was at this time that my friend Robert Lancaster of the N.C.B. asked if we could think of a way to support the roof of a coal mine by a support which did not need removal and setting as the coal face advanced. The result was what we described as a walking chock. Although this design was not of the type that went into production it was the first effort to try and meet Robert Lancaster's forward ideas.

In 1955 both Mr. Banbury, my Company Secretary, and Mr. Lionel Barber who had for long been our financial consultant, joined our board. This was the year when two honours were to come my way. The Royal Aeronautical Society awarded me their gold medal for 'Outstanding Achievements in the Design and Development of Aircraft Equipment. '

On the 10th January I was greatly honoured by being admitted an Honorary Freeman of the Borough of Cheltenham, the first to come from industry. The citation read, 'In recognition of the distinguished and eminent services rendered by him in the field of industry and to the progress and life of the town over many years and in token of the esteem and affection in which he is held. '

It was said by the speakers that my success story was Cheltenham's success story for since the start of my business in 1931 the population had doubled. I recalled that when I came to Cheltenham thirty years previously I was a young engineer poor in pocket but not without ideas. It was fortunate that these had proved successful. In mentioning that

George with wife Marguerite and sister Jesse Fell at Buckingham Palace
investiture

I was the first industrialist to be so honoured I said this reflected the changing pattern of life in Cheltenham.

Cheltenham Town Hall was filled with many distinguished visitors including the Speaker of the House of Commons, Mr. Morrison (later Lord Dunrossil).

1956 was also an eventful year. Mr. Hunt who joined as a young man from Cheltenham Grammar School in 1935 was elected to the board.

It was on my birthday, the 27th April, I received a letter from the prime minister saying that subject to my agreement, he could recommend Her Majesty to confer on me the honour of a knighthood. Quite a birthday present! This was something I had never sought and I was honoured to think that my work had been considered such to merit this recognition.

Having built up my business from nothing and brought considerable employment to the west of England there was of course, the substantial contribution I had made to the war effort. I was president of the Royal Aeronautical Society in Coronation Year when Her Majesty was patron of that society, and at that time I had been chairman of the North Gloucestershire Disablement Committee for some years and a director of Remploy.

On July 12th, I attended Buckingham Palace to receive the accolade and was accompanied by my wife and my sister Mrs. Fell.

The Palace ceremony was impressive and amongst those knighted at this time were my friends, Sir Alfred Pugsley, Sir Leonard Hutton and Sir Graham Rowlandson. Others were Sir Solly (now Lord) Zuckerman and Sir Eric Ashby. In the evening I gave a celebration dinner party at the Savoy Hotel.

I well remember the morning this honour was announced. My daughter then aged four, said, 'Daddy, why does the telephone never stop ringing. "People are congratulating me. ' 'But for what are they congratulating you Daddy?' To which I replied, 'The Queen has given me a present. ' 'Can I see it?' my daughter continued. 'I am sorry' I said, 'It is something that cannot be seen. ' 'Well,' she replied, 'I don't think much of that as a present!'

As I have frequently said the whole of our country's economy is based on industry, primarily those who produce products for export, but looking at the manner in which honours are given one would not suspect this.

In the Birthday Honours of May 1956, 32 Knight Bachelors were created yet none were given for services to Industry, save remotely one for services to Shipping. Even my knighthood was for my services to the disabled.

Knighthoods were mainly for political and public services, to civil servants and sportsmen. This is due, no doubt, to honours being recommended to the Queen by the Prime Minister with advice from senior civil servants.

On the 15th of June we celebrated our silver jubilee as a company when the Rt. Honorable Reginald Maudling, M.P., Minister of Supply, and my old friend, the late Sir Sydney Camm, spoke at the luncheon. The minister said he was gratified to be asked to propose the health of a firm connected with the aircraft industry because it was well known that the main handicap the aircraft industry had was the Ministry of Supply.

A rare event in modern history also took place that year when the employees of Dowty Group presented me with a cheque for over £2,000, collected as an expression of their esteem and gratitude for whom they termed a 'model employer'. I handed this sum to the Old People's Welfare Committee who were pleased to name their headquarters Dowty House. I was deeply touched by this expression of friendship and goodwill and at that time I said I wondered how many employers would have been so treated.

I then bought 93 Church Street, Tewkesbury and after making structural alterations and decorations gave that house, rent free, to the Tewkesbury Old People's Club as I felt that since our businesses were so firmly established in both towns Tewkesbury ought not to be overlooked.

My old friend Sir Roy Fedden was an ardent fisherman and invited me and my wife to Scotland as his guests. Although not a fisherman myself I am always interested in another man's sport. We stayed with Sir Roy at the Belgrave Arms at Helmsdale. The room we were given had a sloping floor, one window overlooked a house with its roof missing and from the other window the view was a concrete wall. Sir Roy was always up early away fishing and so active he ran from pool to pool. Sandwich lunches were taken in a hut that stank of fish.

In the late afternoon we repaired to the hotel. My host neither drank, smoked or played games. The evenings were spent looking at the records of the river and admiring boxes of flies. Sunday is a non-day in Scotland, no fishing, so for a diversion we were taken over the hills to a cottage to meet the woman who made the flies.

When we left the Belgrave Arms for home the hotel proprietor ran after the car shouting and waving his arms. I thought I must have left something behind. He came up puffing and blowing to present the bill for my stay!

I knew Sir Roy Fedden nearly all my working life. He was a brilliant aero engine designer although very opinionated and a technical autocrat.

On two occasions I was on board ship when he was leading teams visiting America and I saw how he ruled his men with a rod of iron. After a punctual breakfast, down to their cabins for work and they only appeared for lunch, then back again until the evening when they were marshalled to the ship's gymnasium for exercise. After dinner, off to bed; and how he kept his eyes open to see that nobody escaped this strict discipline.

Sometimes he would ask to see me on the pretext that he wanted advice. When he arrived he told me about some project or problem and what he proposed to do about it. He never asked me a question or indeed gave me the opportunity to make comment. He would then get up and thank me profusely for my advice and depart.

On one occasion he was holding a meeting with his senior men discussing a solution to some problem. When he had propounded his views he went round the table to ask what each thought of his proposals. When they agreed all was well but when there was one so imprudent as to suggest some other way, Sir Roy banged the table saying, 'What, can I get no co-operation?'

In April 1956, the late Mr. W.S. Morrison (created Lord Dunrossil) the then Speaker of the House of Commons and member for Cirencester and Tewkesbury, formally opened the large factory that had been built at Ashchurch for the mass production of hydraulic pit props, with its automatic multi-welding machines and conveyors for modern flow production. At its outset the plant consumed 15 miles of steel tube each week and the following year, in 1957, we could claim that the millionth Dowty pit prop had come off the lines. Improvements were continually being made to this product but by intensive care the product was continually cheapened so that the selling price of the props remained constant. Licences for the manufacture of the props were granted to firms in Germany, France and Japan.

The spectacular success of the Dowty pit prop induced the National Coal Board to throw down a new challenge when they asked for a hydraulic roof support system which could bring a high degree of automation to the process of coal getting. This challenge was enthusiastically taken up and by resolute research and development the difficulties were overcome, and the outcome was the Dowty Roofmaster, the world's first semi-automatic system of roof support.

As the coal cutter moved along the face so each hydraulic unit of the Roofmaster installation lowered itself from the roof, advanced forward carrying with it the conveyor and then reset itself against the roof.

This was the first stage only, however, for later on the hydraulic chock which was a framework with four pit props one at each corner, became the accepted practice, as this unit had much greater stability especially on sloping seams. This enabled the introduction of automation to the mine faces so that the coal could be cut, transported and all operations underground became completely automatic. In much less than 20 years there had been a remarkable revolution in the mining of coal.

The hewers of coal no longer had to wield axe and shovel as they did a few years ago. That drudgery had gone but no-one talks about this. There is always the image of miners with dirty faces, the public believing that mining is the arduous job it was. This is the age of mechanisation when coal is produced by push button methods. But it seems that the greater the outlay on expensive machines and mechanisation, the less is the output. Of course, conditions underground are not pleasant but neither are conditions for those who work out of doors in all weathers.

The following year another company was formed, D. G. Packing. With the escalation of exports specialised packaging became very important.

In July 1956 I was invited by Lord Thornycroft, president of the Board of Trade to join his advisory committee on exports. Monthly meetings were held, the chair being taken by the president. I attended many of the meetings until 1959 and during most of that time the president was Lord Eccles. It was of interest even at that time to note the lengthy discussions on the common market. The president said the arrangements were more sinister than had been expected while Reginald Maudling, Paymaster General, was much concerned with the Treaty of Rome over which he seemed not too enthusiastic. I found Reggie was a most amiable and likeable man.

A member of the committee at that time was Sir William Rootes, later Lord Rootes, who said the government must avoid being jockeyed into making substantial concessions to the French. It seems strange that nearly 20 years later we are still having these problems.

Before attending the last meeting of the committee I heard of the passing of my friend Sir Claude Gibb, a most capable man with whom I had a close association.

At that time Freddie Errol (later Lord Errol) was parliamentary secretary to the Board of Trade. I always warmed to him as a member of the government for he was a trained engineer.

During September and October 1957 I attended the sixth Commonwealth Mining and Metallurgical Congress in Canada. Starting from Vancouver I visited White Horse, Dawson City, Uranium City, Kimberley, Edmonton, Saskatoon, Winnipeg, Sudbury, Timmins and Toronto, where I left the tour. We travelled in the longest train ever seen on Canadian railways which transported us by night and stopped during the day for visits to the various mining operations. During this trip we saw the mining of lead, zinc, iron, tin, nickel, gold, potash, uranium, oil and gas. It was an impressive tour but highlighted the little use made of British equipment in these operations. At each port of call we were greeted by a brass band and the local dignitaries.

I spent two days in Vancouver and visited manufacturing plants and also Vancouver University. The University buildings are excellent and in a fine location and I was impressed with the facilities. Mr. Raymond, a master from Cheltenham College, covered part of the tour and visited part of the university and made his comments to me on the standard of education which he found. A boy from Cheltenham College who could never have entered a British university through his lack of ability, was accepted by Vancouver University and was considered a satisfactory student. Mr. Howard, son of Lord Strathcona, who was educated at McGill, told me he could never have gained his degree at Cambridge.

Mr. Diamond, the President of the Congress, stressed in his opening remarks that there was a grave shortage of technicians in Canada. On speaking of this matter to him, it was apparent Canada is only doing a fraction of what it should in training technical people. They seem to rely on getting young executives who have been trained either in Britain or other European countries. I was impressed by the number of university graduates with no better jobs than that of Salesmen.

With all the forests of timber in Canada I was amazed to hear that a thousand people had been laid off from the lumber industry and that in British Columbia it was cheaper to import timber than use the natural product.

The next stop on this trip was Trail, British Columbia. This was a town of 14,000 inhabitants in a delightful setting, perhaps spoilt by the old and dilapidated buildings of Consolidated Mining and Smelting Company of Canada, who run the town. Like many of these out of way places they have to cater for the social amenities and have such things as a skating rink to hold 5,000 people. There are also eight carting rinks and golf courses. I found unusual working conditions here in that people worked 12 days on, four days off.

The following day I reached Kimberley and was entertained again by the same company as that we visited in Trail but here we had the opportunity of visiting the fabulous Sullivan Mine, producing lead, silver, gold, antimony and bismuth, zinc, cadmium and indium. Much chemical fertiliser is also produced from the by-products.

Kimberley is a most delightful and attractive town in a wonderful setting with charming and hospitable people, and this was certainly a highlight of my trip. From here I visited the Crows Nest Pass which was a coal mining area. We stayed at a motel in the mountains in a most magnificent setting.

About two miles east of the Crows Nest Pass a road climbs for a distance of approximately eight miles to the top of Tent mountain, nearly 8,000 feet above sea level. This is in effect a mountain of coal astride the Alberta/British Columbia border. It is a most spectacular operation carried high up in the mountains, from which there are fantastic views that can well be described as 'mining in the skies'. Here there are millions of tons of coal forming a plateau about 500 yards wide with a depth of several hundred feet. Coal is to be had for the mere picking of it from the surface and transporting it down the mountain.

At the end of the day we were entertained at a motel called the Turtle Mountain Playground. The only evidence I had of the playground was an empty swimming pool! The great interest in this locality is the famous Frank Slide. It was here in April 1933 in the early morning that a whole section of Turtle Mountain, some 70 million tons, broke away destroying the complete township of Frank. It is an awe-inspiring sight, remarkable how rocks the size of cathedrals travelled two miles across the valley. The town was completely wiped out.

My next stop was at Edmonton to visit the oil wells and the natural gas industry. The economy of Alberta is stupendous. There are three strata of the ground, the surface growing wheat, the next layer producing gas and further down, the oil-producing layer. And all are operating profitably. I noticed that the pumps were stopped in most places, and I was told, only worked for two hours out of twenty-four since they were over producing for the demand and yet in spite of this exploration continued at a high rate, to discover further fields.

My next stop was at Saskatoon where the main activity was potash mining. The following evening we reached Winnipeg and on the following day I was invited to speak to the Empire Club and was kindly received by the Lieutenant Governor of Manitoba.

The next port of call was Atikokan to see a really fantastic development. Some 40 years ago Canada lacked two basic essentials for an industrial civilization, iron ore and crude petroleum. The knowledge that large deposits of iron ore were available at Steep Rock, Atikokan, pointed the way to the solution of the first problem. Although small deposits have been found from time to time it was impossible to locate the main body until a few years previously, when it was found to lie under the bed of the Steep Rock Lake. In order to make these deposits available it was necessary to empty the Steep Rock Lake and divert the Seine River. This was a gigantic undertaking involving as much work as the building of the Panama Canal.

The next area visited was Aldoma uranium area at Blind River. The ore body has outcrops at two places seven miles distant and the strata dip to a depth of 4,000 feet. The conditions at Blind River when I was there were extremely bad from the point of view of labour, the nearest township being Elliot Lake where people lived in caravans with no proper sanitation or water facilities, and the ground around these camps was a sea of mud. The 11 miles' travel to the mine was along a tortuous road so bad that there are wrecks of automobiles every few yards along the unpaved highway. All this is due to the need to start production in a great hurry but, as I have said, the conditions everywhere were most unsatisfactory.

My next visit was to Sudbury and Falconbridge to see the International Nickel Company with a fine operation.

Travelling 600 miles north I expected to find bleak and uninteresting country. On the contrary I found the area quite charming. This was a visit to Timmins to see the gold mining. It is claimed that this is the third largest gold producing area in the world. Like all other areas I was taken underground but, of course, gold mining is quite different from all other types of mining, I found it much more elementary with dependence on heavy timbers, most of which did not look efficient or safe.

An amusing incident occurred to me along the trip when my wife left the train to spend some days with her mother who lived in Canada. Laundry was a problem and I was told about the liquid I should put in the wash basin. So I put in what I thought was a liberal amount and then the fun began. The small compartment became full of soap bubbles. The more water I ran the more the bubbles multiplied. I put handfuls down the toilet with no apparent effect, I opened the window and tried throwing the stuff out, I don't think I have been so exasperated before. How I ever got clear of the mess I cannot remember when I recounted this experience to my wife all she could do was to laugh and to say, 'But surely you should have known to use only one or two drops.' Well how should I when I had never used the stuff before.

In May 1957 we held a luncheon at Ashchurch where the director general of production for the N.C.B., Mr. Henry Longden, and many officers of the N.C.B., were present to celebrate the manufacture of the one millionth hydraulic pit prop. At that time we could boast of exports of these props to Australia, Hungary, Germany, Belgium, France, Czechoslovakia, South Africa and Holland. Some years later we completed the manufacture of over ten million of these props before advancing roof supports became the order of the day.

Alec Mills and I visited Germany to negotiate a licence for the manufacture of pit props with the Salzgitter Company of Dusseldorf. I met Alec Mills on Liverpool Street Station and he was resplendent in a bowler hat, I suppose to indicate his nationality.

We had rather fearsome negotiations with the Germans who thumped the table and were domineering and rude. I remember I gathered up my papers, pushed my chair back, wished the meeting good day, and made for the door. Of course, this was not what the Germans expected and as I suspected I was not allowed to go. There was a change in their attitude and finally an agreement was signed.

Some months later when their pit prop factory was in operation my wife and I were invited to visit Salzgitter for the gala opening. We were royally entertained with a banquet at Bonn attended by the German Minister of Finance and other notables. The following day we were taken by car, 80 miles down the Rhine, to a large yacht flying the flags of Germany and the Union Jack. On board was an orchestra to play the songs of the vineyards while we were to drink wine from the vineyards as we passed them by. We were told the skipper of the yacht held the Blue Riband of the Rhine so was acknowledged as a most accomplished pilot.

Those who know the Rhine are aware of the numerous barges on that river which are towed by a tug that always seems a long way ahead of the barge.

With all aboard the yacht made its way into the river but incredibly the pilot crossed between a tug and its barges. A great scream went up, the wire hawzer towing the barge rose up over the yacht, people fell flat and were scattered over the deck. An incredible sight with much blood flowing. I was on the steps of the cabin into which I took a flying leap – the yacht was dismasted. By the grace of God the yacht did not capsize as well it might have done. We put back to shore and ambulances took the injured to hospital.

The river police were much in evidence and the pilot taken away for questioning. I suppose he was prosecuted. I never did hear the sequel of this near tragedy and quite stupid accident. Two hours later the ambulances returned with those able to continue but on a trip which was much curtailed and of course, the day was ruined.

About this time the company I had set up in the Isle of Man ran into difficulties. Short of work, following the cessation of wartime orders, they had taken on an extensive programme for a spinning or weaving device called a Pern Winder. Extensive stocks and work in progress accumulated but the customer did not pay and went bankrupt, placing the company in jeopardy. The banks were proving difficult.

To see what could be done I remember flying with that grand old pilot, Captain Olley, with Mr. Lionel Barber, to the island one Sunday morning in foul weather. We were flying but a few feet above the ocean but even so we missed the island, for after what seemed an interminable time we turned back and eventually found Ronaldsway.

I am pleased to say we were able to pacify the bankers, and after a year or so were able to get the company back on its feet.

I found it extraordinary how a large organisation with original ideas used underhand methods to take advantage of another company's developments. A company which is one of the largest suppliers of materials for the engineering industry, had access to our designs and programmes when making quotations for the supply of forgings and other fabrications. In the space of a few months we found them competing with us for the manufacture of pit props, railway buffers, railway wagon couplers and roof seals. This breach of confidence was so flagrant we forbade any member of that company to visit any of our companies, and we placed no orders with them for many years.

Another experience with an internationally known American company shows how ideas are wilfully stolen. Our engineers visited this company to show them certain developments for which we had applied for patents. Some years later we found that following this visit the American company had applied for an American patent covering the same idea. When confronted with this sharp practice the American company claimed this was an independent development by their engineers. This is further evidence of the ruthlessness of American business.

[A handwritten page of notes can be found here, probably left in by accident]

The Business Expands

I N 1957 I received an invitation from the president of the Board of Trade to become a member of his private committee. This was the time when Reggie Maudling was Paymaster General and was concerned with the Treaty of Rome. Much of the time of this committee was given to a review of the Treaty terms. Of the six or seven members of that committee most were against entry into the E.E.C. and I have always been anti-common market. Our entry was advocated by those whose belief in the advantages only appeared to be wishful thinking. The European industrialists to whom I spoke welcomed our entry as it would give them a larger market. This was exactly what our politicians were saying but all could not win a larger market. We could only sell more into Europe if we had products Europe wanted, but joining the common market would not produce these goods. Indeed three years after joining the E.E.C. the U.K. trade deficit was £1, 800 millions a year – ten times the figure for the last full year before joining the community.

In 1957 Sir Godfrey Ince, at one time permanent secretary to the Ministry of Labour, joined our board. I had known Sir Godfrey for many years through my association with Remploy and other matters. His work in the civil service was impressive and I was pleased to have a wise and helpful counsellor. Among his other appointments he was chairman of Cable and Wireless. Although suffering ill health he never let up on a life of great activity. Sir Godfrey Ince was a most charming man but unfortunately had domestic problems so was grateful for the help my wife was able to give him in such matters as personal shopping. He suffered with diabetes and heart trouble but would not give up his very active life, although we knew he should not have undertaken long overseas trips on behalf of Cable and Wireless of which he was chairman.

He died in December 1960. It was then I invited Sir Miles Thomas, now Lord Thomas of Remenham, that well known figure who had been Lord Nuffield's right hand man and then chairman of B.O.A.C. , to join my board.

This year Innsworth Metals, a company I founded in 1950, became part of the group.

In 1958 we were interested in the development of hydrostatic transmissions and Sir Roy Fedden said he would visit Professor Hans Thoma in Germany, who was a recognised authority on these products. The upshot of this visit was a statement by Professor Thoma that agreement with German companies precluded him from assisting us but he proposed that his son Oswald could do so.

Oswald was a good engineer but excitable, nevertheless he was helpful in our designs. About 1963 we found that Professor Thoma had another son, Jean, who was helping another British company on similar work. As the brothers Jean and Oswald frequently met we thought it was a most unsatisfactory arrangement for there was no doubt that our development work could be so easily transmitted to a competitor.

On this account there was a period when Oswald ceased his work for us but after the elapse of some time we managed to sort our difficulties and come to a different arrangement. Unfortunately this was short lived as Oswald died of a stroke.

In October 1958 Lord Hives, chairman of Rolls Royce, visited me concerning the Rotol and British Messier companies, jointly owned by Rolls Royce and Bristol Aeroplane. These factories established during wartime were located near Staverton Airport and two miles from our headquarters at Arle Court. The principle product of the Rotol company was the V.P. propeller, originally developed by the Gloster Aircraft Company when I was in their employ. The business of British Messier (originally an offshoot of the French Messier Company) manufactured aircraft landing gears. They were brought into this country at the commencement of the war at the request of Sir Frederick Handley-Page who believed that no-one in the U.K. could produce what he wanted. Ironically at the height of war production, the Handley Page 'Halifax' bomber was halted for lack of undercarriages and Lord Beaverbrook sent Sir Robert Renwick to ask me to take over production of Halifax undercarriages to my design. Part of the Messier problem was the introduction of metric dimensions which caused intolerable difficulties and delays.

At the time Lord Hives visited me British Messier were making undercarriages for the Bristol 'Britannia', the Folland 'Gnat' and English Electric 'Lightning', work obtained through the influence of Rolls Royce or the Bristol Aeroplane companies.

Lord Hives came quickly to the point and said that Rolls Royce and Bristol Aeroplane were not happy bedfellows and would I take over the company? To this we agreed after writing down what was obviously gross overstocking. When the deal was consummated Rolls Royce and Bristol Aeroplane presented us with a fine sports pavilion.

Taking over a sizeable company employing 4,000 people did not prove easy – the employees were suspicious of every move we made. The problems we found concerned top management. These appointments had been made by Sir Reginald Verdon Smith, who had a poor reputation for picking men. The chairman was Lieutenant-Colonel Sir John Evetts and his works director, Colonel King, both charming men with military backgrounds but lacking industrial experience. It was soon clear that change in the administration was imperative. No-one could report factually on the position of orders, promises were meaningless and troubles seemed everywhere. I am sure the top management knew of their inadequacy.

We amalgamated Dowty Equipment with the Rotol and Messier companies, calling the new company Dowty Rotol Limited but it took several years for the combined organisation to settle down. Nevertheless by their acquisition we became Europe's largest manufacturer of aircraft equipment.

In 1959 Dowty Group Services was formed to take over many activities common to all our companies, public relations, advertising, patents, works and buildings, education and so on.

In that year I addressed the London School of Economics and my opening and closing statements are worth recording. 'Nowadays it is widely assumed that without strong financial backing few new major businesses can emerge and succeed; that new products are almost invariably the results of the research work of expensive teams of scientists and technologists; that the location of factories is the outcome of exhaustive enquiries into the relative advantages of here or there; and so on. The Dowty Group does not fit into this pattern at all. For example, the location of our businesses has been entirely fortuitous. When I started single-handed in 1931 I happened to be working with the Gloster Aircraft Company and since I did not have the money to move elsewhere I started my business where I was.' I wound up saying, 'Finally I have found that money in any business is not the first requirement – good ideas are by far the most important commodity'.

The year 1960 saw the formation of yet another company, Dowty Marine, with the late Donald Campbell as managing director. Donald was a great and lovable character but everything he did was on a lavish scale. He took many risks with his speed records on land and water. We were all very sad when he lost his life attempting a water speed record on Lake Coniston.

Dowty Marine was formed to manufacture a jet boat propelled by a water jet to the design of Mr. Hamilton (now Sir William) of New Zealand. Sir Roy Fedden visited New Zealand to appraise the jet propelled boat and it was due to his enthusiasm for this form of propulsion that we took a licence to build this boat. This business was not financially successful although the jet unit was successfully applied to life boats and the military amphibious craft, the Alvis 'Stalwart'. I took out patents in 1959 to provide improved steering of this boat when the jet gave reverse thrust.

In the early autumn of 1959 on a trip to America I visited Indianapolis to see a company, Indiana Gear. Their president, John Buehler, met me at the airport with a helicopter he was piloting. Taking off we flew over the city and landed on the roof of his main factory. This was a comparatively small operation employing some 500 people, manufacturing gear heads for helicopters and there was another section making jet propelled boats similar to our Turbo-craft. I was not impressed with his shops which I found poorly organised. With such a small company it was surprising they could operate a DC.3 aeroplane and a Sikorsky helicopter.

At lunch time we took off by helicopter which Buehler landed at a restaurant in a shopping plaza. After lunch we took off again for a lake where the boats were tested and after trying out various models we flew over the surrounding countryside and over Buehler's yacht club where I had a most extraordinary experience. There was a sailing yacht on the lake which we flew over. Banking the helicopter we saw the yacht heel over under the down draught from the helicopter, the sails resting on the water. The crew, which we later discovered consisted of three ladies, were all in the water and from afar we watched their rescue by boats from the club. The Police visited Buehler that evening for a statement but I do not know the outcome of that unnecessary mishap caused by his ego and showmanship.

Buehler was most hospitable, unnecessarily so, and carried things too far. That night he had a cocktail party for all his work people, presumably to meet me, and how they all drank! I could not help feeling that here was a small company operating on a very lavish scale.

During that same trip I visited Milwaukee to hear one of my technical men read a paper on hydrostatic drives. I then went to Chicago and lunched with Borg Warner directors which included Robert Ingersoll, their president, and Mr. Harry Whittingham who has become a great friend. This was a time Borg Warner were taking legal action against us over the use of pressure-loaded side plates on gear pump. We knew they had no case and this was the situation finally established.

International Harvester were interested in taking up a licence for hydrostatic transmissions and I dined with Frank Jinks, president, and his other vice presidents. The draft agreement read more like a novel, the work of too many lawyers. I found it difficult dealing with American companies. Our English methods are more straightforward and simple. I once heard an American say, 'What can we do to make it legal?' and that, I think, indicates their philosophy.

I always enjoyed my visits to Precision Rubber Products of Dayton and to stay with Bob Allen and his charming wife, Connie. Many years ago the founder of that company, the late Jack Taylor, took out a manufacturing licence for bonded seals and since then sales of this product have been considerable. On this visit I flew with Bob Allen in a light aeroplane to a new factory they had opened in Lebanon, Tennessee, some 300 miles south of Dayton. Although he did not admit it I am sure this factory was chosen for its location in a much lower labour cost area. The factory is set in 30 acres of land and Bob Allen claimed to be the most up-to-date rubber plant in America. Certainly I thought it was first class and was most impressed. I found it interesting that the civic dignitaries of this small town of Lebanon were at the airport to greet us.

From here I travelled to Williamsburg in Virginia, a wonderful historic town I had heard so much about and the landing place of the Pilgrim Fathers. It was quite delightful. The town was kept as it was some hundreds of years ago. Every shopkeeper was to be seen plying his trade, the apothecary with his pestle and mortar, the cobbler at his last, the printer at his press and so on. The hotel staffs were dressed in period costume and the menus were of dishes of by-gone days.

The architecture of the shops, the homes and the large estates, are all those of long ago beautifully preserved.

The William and Mary College is a magnificent building, a fine seat of learning which often takes students from the U.K.

While walking through the streets of this town I met a director of the Bristol Aeroplane Company who some months previously had negotiated the sale of the Rotol company.

A most interesting experience was being flown over New York by helicopter. Seeing this great city from the air gives one an entirely different concept of Manhattan. I was surprised at the amount of open spaces in what always appeared to me to be a congested city.

The amount of river traffic was a revelation with many boats plying between Manhattan and New Jersey.

It was interesting to see congestion on some roads and the usefulness that the helicopter service must have for controlling traffic flow.

The flight took me to the Statue of Liberty and the pilot came so close I feared the blades would hit the statue. I felt I could have lent and touched it.

There cannot have been so many changes in ministerial appointments as that of Minister of Aviation. The case of Duncan Sandys was no exception. Appointed in October 1959 he was replaced nine months later in July 1960 by Julian Amery.

It was during his reign as Minister of Defence that Duncan Sandys made the most disastrous decision – there were to be no more fighter aircraft built in the U.K. but henceforth the country would depend on guided missiles for defence. Indeed the Hawker 'P.11 21', then three-quarters complete, was placed on the scrap heap. At a stroke we lost our overseas fighter aircraft business and handed it on a plate to the Americans. Twelve years later we were buying 'Phantom' fighters from the United States and in 1972 entered a consortium with the Germans to produce yet another fighter, the M.R.C.A. !

Although Duncan Sandys takes the responsibility for this failure the man regarded as making this decision was Solly Zuckerman (now Lord Zuckerman) then honorary secretary for the Zoological Society! His books on the love life of the ape obviously qualified him for the position of chief scientific advisor to the Secretary of State for Defence! Industry regarded these decisions as nonsense but nothing we could say had any effect.

There was a considerable move towards the development of guided missiles. I was invited to take a major interest in this work but declined. There were too many competitors for orders from a single customer and I preferred to develop products I could sell worldwide without let

Sir George greeting Her Majesty the Queen Mother in 1960

or hindrance. As it transpired several companies built up large activities which were eventually axed overnight, so their efforts, which could have been better directed, proved abortive.

Julian Amery, the Minister of Aviation from 1960 to 1962, and B.O.A.C. were responsible for causing prospective customers throughout the world to believe that splendid aeroplane the V.C. 10 was a failure. Their attitude was quite incomprehensible.

For some years I had been interested in the industrial development of Malta, having been invited by Dom Mintoff, their prime minister, to serve on a committee under the chairmanship of Lord Hives. In 1960 when Lord Hives resigned I was invited by the Secretary of State for the Colonies to become chairman, in which capacity I served for three years until our committee resigned en-bloc being unable to get any decision or attention to our requests by the then prime minister, Borg Olivier.

The years I spent helping Malta were interesting and I always found Dom Mintoff a dynamic person who got things done. Sir Maurice Dorman, the governor general in later years, was a charming and most helpful man.

In 1961 I set up a company in Malta. It was apparent that many industrialists who came to Malta did not introduce the right type of businesses. Malta's need was labour-intensive factories to mop up the large numbers of unemployed. Furthermore, the products needed were those which could be transported economically by air since other forms of transportation were almost non-existent. Unfortunately, in haste to get companies into Malta many wrong decisions were made. As a committee member I objected to certain applications such as those of Riggs Welt and Rambler Cars. None of these applications made sense to me and so they proved to be. Large grants were given to companies which never produced anything or employed anybody! Help should only have been given to those who perform. The factory we started to manufacture rubber products in is now the largest factory in Malta, employing 1100 people, and is making 20 million rubber seals each week.

As chairman of the Industrial Development Board I was given certain instructions to observe whenever I arrived on the island. The first was to visit the Roman Catholic Archbishop and sign his book. I found it extraordinary that I was not expected to act in a similar way to the Lt. Governor or yet the prime minister. The Roman Catholic Church is all powerful in Malta to the extent that it is illegal for any clergyman other than a Roman Catholic to wear a 'dog collar'.

In the Museum at Valetta I was interested to find one of the Gloster 'Gladiators', that famous trio 'Faith', 'Hope' and 'Charity' which were the main defence of the island at one period of the war. This old aeroplane with our internally sprung wheels has recently been restored.

Sir Roy went to Austria to look at the Wankel engine for me but there appeared too many problems and we both agreed that the development costs would prove too costly for our company. As things have turned out that decision was right.

In 1961 my wife and I visited Sweden on the occasion of British Week in Stockholm where my company was showing products at an exhibition. The Swedes had always been our good customers and I was honoured that King Gustav spent some time talking with me about my work and our business in Sweden. My wife and I were invited as his guests to a Royal Command performance of Benjamin Britten's 'Midsummer Night's Dream'. I am no lover of modern music and found the occasion most depressing. Under other circumstances I would have left the performance but I could not do so out of respect for His Majesty. I sat in agony listening to the discordant sounds and a performance in a language I could not understand.

Dowty equipped aeroplanes

For many years since the mid-thirties I had taken a great interest in the Society of British Aircraft Constructors, having been chairman of the Equipment Group Committee and a member of Council. During my early years on the council I was instrumental in bringing about several changes. Associates were called 'A' and 'B' members. The 'B' members being the equipment suppliers! I was responsible for the change to 'Material Suppliers' and 'Equipment Manufacturers' – more dignified descriptions. Later on, under some pressure, I got the council to agree that the larger equipment suppliers should become full members – a status previously reserved to the engine and airframe manufacturers.

I was elected their president for 1960-61, the only person at that time to have held the presidencies of both the technical and business associations of the aircraft industry. From 1961 until 1969 I was treasurer of the society.

During my office as president I entertained among others the president of the Board of Trade, Reginald Maudling; the secretary of state for war, John Profumo; the Chairman of B.O.A.C., Sir Matthew Slattery; the Minister of Aviation, Duncan Sandys; and the minister of defence, Harold Watkinson.

Each year the Farnborough Air Show is regarded as the shop window of the aircraft industry and attracts hundreds of guests from overseas, from foreign governments and the airlines. On the open days many hundreds of thousands of the public come to see the show. In my presidential year it was the usual impressive affair and I entertained the prime minister, cabinet ministers and a host of important guests.

Squadron Leader Neville Duke, the well known Hawker Chief Test Pilot and now my personal pilot, flew me and my family into Farnborough, a privilege accorded only to the president. It is the heaviest week for the president starting on a Sunday it finishes the following Sunday when the pilots, officials and press are entertained to thank them for their contributions towards the success of the show.

At our annual dinner I mentioned that I was the first president not to be connected with an airframe or an aero engine constructor. The principal guest at that dinner was Lord Carrington.

In 1963, after the death of my brother, Robert, in the Isle of Man, Iloman Engineering became a part of the Dowty Group.

In 1964 I was invited to become chairman of the Engineering Industrial Training Board but when I found it would require my services for four days a week I had no alternative but to decline and I was pleased that my old friend Sir Arnold Lindley took on this job.

In 1964, Dowty G.m.b.h. was formed and later on our activities in Germany were concentrated on the sale of rubber seals. We joined with the Klockner company in this enterprise and operate under a company called Dowty Klockner. From a modest start this business has grown into one of considerable size and imports most of its products from our Malta factories.

In 1964 I was to receive another honour when the Borough of Tewkesbury elected me an Honorary Freeman. This was a splendid occasion with the Duke of Beaufort, the Lord Lieutenant of the County, present and presiding at the dinner that followed. On that occasion I was pleased to present to the borough something they had never possessed – a coat of arms.

The crest shows a lion holding between its paws a black pear which was taken from my own arms. The historic town of Tewkesbury had been in decline for over a century but the coming of my factories had given the town a shot in the arm so to speak, and in the twenty years since setting up these factories the population more than doubled. Like Cheltenham, I was the first industrialist to be so honoured by this ancient borough.

On this occasion I recalled that many industries started and foundered in Tewkesbury over the years. The coming of the railways, instead of bringing industry to the town, proved a social and economic disaster. When I set up my factories in Tewkesbury the population was the lowest it had been for 150 years – only some 4,000 inhabitants. Referring to the manner in which I started industry in Tewkesbury, I said that this was a wartime measure and I was driven in desperation through the unavailability of any other building in North Gloucestershire, to occupy the unwanted forage store on Ashchurch Station. This turned out surprisingly well for we now have considerable factories in the Tewkesbury area, employing thousands of people. So it was through this antiquated railway building that this industrial activity started and the railways, through their unattractive forage store, could yet claim credit for the coming of modern industry but in a way which neither they nor anyone else could have foreseen.

In 1961 Dowty Electrics was founded and another business, Boulton Paul Aircraft, Wolverhampton, was taken over. This came at a time when the aircraft industry was being restructured. Mr. J. D. North, chairman and managing director of that company, told me he was being threatened with a takeover from Staveley Industries. He was anxious this should not take place. On the other hand he would welcome an overture from my company but if we were interested we must move quickly. Within a week we had agreed terms which were approved by Boulton

Paul Aircraft shareholders. As the company had long since given up the manufacture of aircraft the name was subsequently changed to Dowty Boulton Paul and is now a subsidiary of Dowty Rotol. Its main activity is aircraft powered flying controls.

In 1965 I took out a patent for the control of railway wagons in modern marshalling yards by hydraulic means. A wagon with low rolling resistance and aided by a following wind could make violent contact with other wagons resulting in severe damage, but if the converse conditions obtain the wagon stopped short of its destination. Because of these variable factors the resulting train did not consist of a compact set of wagons so that these were brought together by means of a shunting locomotive. Our wagon control system overcame these difficulties by using a series of hydraulic units placed on the inside of the running rails in such a way that their pistons were depressed by the passage of the wagon wheels. These units were speed sensitive so that a wagon moving too slowly was accelerated and a wagon going too fast was automatically braked. These systems are in use on two of our country's large marshalling yards where the advantages of this system won a Queen's Award for technical innovation. This system has attracted the attention of railway engineers from all over the world and installations are in operation in several overseas countries.

On January 30th 1965, on the occasion of Sir Winston Churchill's funeral, I was in the Swiss Alps in the village of Wengen. The little English church had provided a number of T.V. sets relaying the service from London. I sat awed through the whole of the impressive proceedings. The spectacle with all its pomp and ceremony was most moving and of such magnificence and good timing I can never hope to see its like again.

In March 1965 I was present at the formal opening of the Dowty Seals factory at Milford Haven by the Rt. Hon. Douglas Jay, president of the Board of Trade. We had moved there at the request of the Ministry of Labour to mop up a pocket of unemployed persons.

In May of this year a new factory was erected on the Isle of Man to replace the old buildings. It was opened by the Lt. Governor Sir Ronald Garvey and on this occasion my wife and I were guests at Government House.

In 1965 I paid my first visit to Japan via Hong Kong. As India was at war with Pakistan I travelled via Bombay where the plane was subject to a long delay over the issue of fuel. Conditions there were chaotic.

We arrived late at Hong Kong after the ferries had stopped running, but the management of the Mandarin Hotel had laid on water transport and we were welcomed by Mr. Hudson, one-time of the Savoy, who we had known for many years.

My time in Hong Kong was spent visiting factories. The larger ones employed thousands of Chinese girls working a 60 hour week for the equivalent of £1 a week. Wages have since risen I understand!

I received a great welcome in Japan. The British Embassy put out an announcement to the effect that my business started as a result of an order from the Kawasaki Company. It was noted in the press that although I was accompanied by my wife I had been too busy to go sightseeing with her! I was entertained one weekend in the country in a hotel in the mountains where we were subject to a typhoon and an earthquake in one evening. The following day the railways were closed to permit examination of the permanent way for damage. Our trip was amended and we were taken to the Pearl Fisheries at Toba.

Visiting one of our licensees in Gifu, a town of 350,000 population, I was entertained at the local golf club where the lady caddies were dressed in kimonos and carried parasols. A splendid touch of colour.

I travelled to Tokyo in the 120 miles per hour train. I was told that if the train was one hour late the fare would be returned! Tokyo is an interesting city but without a name to a road or a street, or yet a number to a house. Postmen must have phenomenal memories. Industry worked a six day week of 48 hours and office staff worked until 5. 00 p.m. on Saturday.

I was disappointed not to see snow on Mount Fuji and so few flowers in the gardens. When I enquired about the cherry blossom I was told by an Englishman who had lived for years in Japan that he had seen more blossom in one orchard in the vale of Evesham then he had seen in the whole of Japan.

In the Japanese newspaper 'Mainichi Shimbun' of September 20th 1965, I was quoted as saying that it would be ten years before the Concorde was in regular service and that its cost would be many times that initially quoted. These views were founded on my life-long association with the aviation industry. I also felt that Concorde was something our country could not afford.

There are many people who still attribute sharp practices to Japanese businessmen but with forty years of experience of many companies I have nothing but the highest praise for the high standards

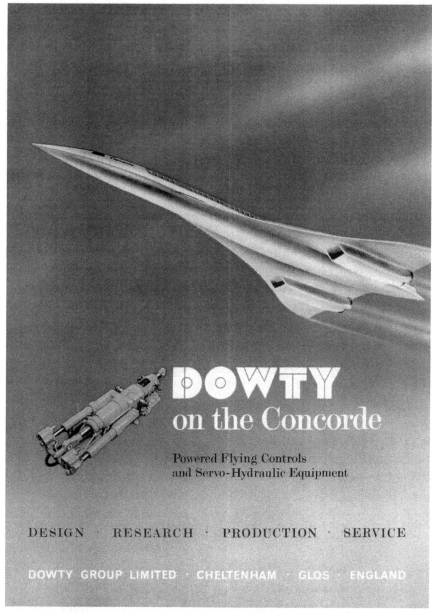

Dowty's involvement with Concorde

and the reliable way the Japanese have carried out their obligations.

They are great givers of presents and I have found this most embarrassing. Whether you visit Japan or the Japanese visit you they always seem armed with presents. Again if I have only met a Japanese once he feels obligated in some way to send me a Christmas card.

Although I do business all over the world I get more greetings cards from Japan than from any other country.

One of the odd things noticed was the Japanese typewriter for typing in Japanese characters. The typist has a tray in front of her carrying some hundreds of characters which she laboriously picked out and put into a frame and then, pressing a treadle, printed a word. That having been done she took the characters out one by one and put them back into their proper place in the tray. The typing of a letter in Japanese characters must take hours.

Women have no part in the social life of Japan, they are invariably left at home. As my wife had accompanied me the Japanese women in expensive kimonos were included in some of our dinner parties but this, I was told, was a rare happening.

Shortly before I returned from this visit my home at Arle Court, Cheltenham, was subject to a break in and a considerable quantity of personal goods and silver were stolen. Although we had a good idea who the thieves were the police were never able to find any of the property.

One evening in March 1966 when dining with Stan and Wendy Walduck at their Hertfordshire home with Ian and Joan Orr-Ewing and Reggie Maudling, my chauffeur sent me a message to say he had heard on the radio that my dear friend Sir Sydney Camm had died that morning while paying golf. Of all aircraft designers of fighter aircraft he stood out alone. Every aircraft he designed was highly successful. To many he was an eccentric but to those who knew him well he was a brilliant man. His great successes were the 'Hurricane', 'Hunter', and 'Harrier'. Here was a man without any technical training who showed once that engineering – even aeronautical engineering – is more an art than a science.

I believe it was because I and Sir James Martin (of ejector seat fame) came up in the same way as Sydney Camm we were such close friends. I feel it a matter of concern that those with a natural flair such as Sir Sydney Camm but without paper qualifications, will find it almost impossible in future to develop their talents as he did. This is the failing of our educational system.

In 1967 on the initiative of Charles Irving, our public relations director, we set up a training workshop in Gloucester Prison, the first of its kind in Britain. We provided machine tools, instructors, materials and work so that prisoners could be trained to carry out useful work hoping that upon their discharge they could become useful citizens.

Looking back over several years since the scheme was initiated we can say it has been most successful. More schemes of this kind should be in operation to send into the world men able to earn a living and contribute to the country's economy.

In 1967 the Royal Aeronautical Society gave me the greatest distinction it can confer – that of Honorary Fellow 'In recognition of many years of outstanding service to aviation'.

John Stonehouse, the Minister of Aviation in 1967, was a most likeable man with whom I had many meetings on my company's business and as an officer of the Society of British Aerospace Companies. It was at this time we had ideas for a variable pitch fan for jet engines and were trying to get support for this development. I thought that Stonehouse would go far for he had great ability. The following year he was postmaster general and he invited me to lunch with him at the General Post Office where he was good enough to give his guests a selection of special issue stamps. He was a charming host.

When the Labour party lost the election he sought a career in business. There are many politicians who out of office have sought to make money in business ventures but lacking experience have made mistakes by association with activities that have done them no good.

Such was the case with John Stonehouse. He set up a company, Export Promotion and Consultancy Services, and came to his friends for support. There were many companies like my own who thought I could help but eventually when no results were forthcoming we had to withdraw our support. Nevertheless, we remained friends and it came as a considerable shock that he ruined his career in the way he did, for it was so unlike the able and accomplished man I once knew.

When the Tory government came into power, Michael Heseltine, the member for Tavistock, was Minister of Aerospace, and during his period as minister I had several meetings with him regarding support for the variable pitch fan project. This matter had been under consideration for several years and although an impressive case had been made for its development, the government had been tardy in agreeing to provide the help necessary which included a considerable modification to an aero-engine and, of course, the eventual ground and air testing.

The minister was most helpful in pressing our case but I had to attend several meetings with him and his technical boffins before it was agreed to award contracts to my company and to Rolls Royce, who were responsible for the engine modifications.

A good friend of mine, Eric Turner, had performed brilliantly at Blackburn Aircraft; coming to that company at the end of the war as chief accountant he quickly rose to managing director and chairman. There was no doubt he had considerable abilities.

One day he came to me for advice, having been offered the chairmanship of B.S.A. At the time the U.K. aircraft companies were being amalgamated and I thought it was a good opportunity for Eric to get into a business that was doing well and had good prospects.

The annual profits of the motor cycle division alone were producing some £4 millions and in 1967 I had discussions about B.S.A. becoming part of the Dowty Group. As it turned out the delays in the production and audited figures and the fact that the managing director of the motor cycle division died at that time, prevented further discussions. Eric Turner became managing director of the motor cycle division but it seemed that it was then things started to go wrong. A highly commercial manufacturing business is a very different undertaking from that of military aircraft.

The dramatic failure of B.S.A. can only be attributed to the management and Eric, good fellow as he was, was not cut out to head such a business. This is yet another case where we have seen an accountant unable to cope with a complex engineering business.

From the earliest days with the Gloucestershire Aircraft Company and specialising as I did on aircraft undercarriages, I was closely associated with the manufacturers of aircraft wheels and tyres and later with wheel bakes. There were two companies manufacturing those products, the Palmer Tyre Company and the Dunlop Rubber Company. The former were unwilling to depart from the spoke wheel construction with fabric side discs, covering the wheel sides, whereas Dunlop were progressive and were the first in the U.K. to offer wheels with metal discs running from the rim to the wheel hub. Later they went to magnesium cast wheels which incorporated the brake drum, and this is the form of construction used to this day.

It was in those early days that I became associated with Joe Wright, a very energetic and progressive salesman for Dunlop aero products. With a young ingenious engineer, Henry Trevaskis, they were mainly responsible for building up a very considerable aircraft wheel and brake business for Dunlop. Joe Wright progressed in the Dunlop organisation becoming managing director of their aviation division until he retired in

1966. He was a most popular figure in aviation circles and throughout all these years we were close friends.

Like other companies manufacturing aircraft specialised equipment, they seemed to be always running late with deliveries but never once was I ever let down. They had the knack of always coming good at the midnight hour.

Joe Wright saw the start of my business and its considerable growth. Our Canada company manufactured their wheels and brakes and in the U.K. our company, Dowty Seals, manufactured their bicycle valves. We had a close and happy relationship.

In 1962 our Canadian company set up a subsidiary in the Bahamas called Ajax International – Ajax being the name of the township in Canada from which they operate. It is through this company that sales to South America take place and there are tax advantages in this arrangement.

In 1968 I was honoured by the Institution of Mechanical Engineers when they invited me to give their James Clayton Lecture on my life's work. I concluded that lecture with the words 'a young man can get on if he has originality and sound ideas and takes advantage of the situation he finds around him'.

In 1968 we took over two further companies, Gloster Engineering and Wandle Rubber, both smallish companies but nevertheless with techniques that we felt were capable of development. Wandle Rubber came to our notice through a curling friend of mine, Remy Ades.

In 1969 we took over the Meco Mining Equipment Company of Worcester, a business whose chief interest is the manufacture of conveyors. This gave us a further interest in mining.

Mr. Paul de la Pena, an old friend of mine and a son-in-law of the secretary of the Meco Company, Mr. Lancaster, came to me one day to enquire if we would be interested in taking over the Meco Company. From this meeting we eventually took over that Worcester company which has proved a successful acquisition and indeed brought me back into my native county. The great character behind the Meco company was Mr. Mark Higgins, a lovable and delightful character and our association with him has been exceedingly pleasant.

It is not without interest in the days when take-over bids are generally accompanied with so much acrimony that all our take-overs, Coventry Precision, Rotol, Boulton Paul Aircraft and Meco, have come about by our being invited to take them over. This I regard as a compliment to our organisation and the way we have presented ourselves.

Later that year another honour came my way when the Queen was pleased to appoint me her Deputy Lieutenant for the County of Gloucester, an honour I was most pleased to accept.

In 1971 the Queen accompanied by Prince Philip visited Tewkesbury on the occasion of the 850th anniversary of the consecration of the Abbey, to present Maundy Money in the Abbey. It was a splendid occasion if somewhat cold. My wife and I had the honour of lunching with the Royal Visitors. I helped escort them with the Duke of Beaufort and Lady Ismay through this historic town. Later that year I lunched with the prime minister, Edward Heath, when he visited Tewkesbury. Peter Walker kindly invited me to accompany the prime minister to the Three Choirs Festival in Worcester Cathedral – a rather dull performance – and afterwards to dine with him at his home in Worcestershire.

On the 4th October 1971, the boroughs of Cheltenham and Tewkesbury held a joint council meeting of Cheltenham Town Hall on the occasion of the 40th anniversary of the Dowty Group, and presented me with an illuminated address and gifts. Happily among the guests attending this ceremony was Mr. Stanley Grove, managing director of Kawasaki, London, the English representative of the parent company in Japan. It was he who in June 1931 signed that order which started me in business. In the evening I entertained to dinner members of the Borough Councils of Cheltenham and Tewkesbury, together with distinguished guests. Reference was made to the one man do-it-yourself concern in the back streets of Cheltenham that had blossomed into a large and flourishing group of companies.

In 1971 my wife and I received an invitation through Lord Thomson of Fleet on behalf of the President of the Lebanon, to visit that country as his guests.

There must have been a great deal of thought and planning put into the trip by our Lebanese hosts for everything was so well executed throughout the Tour. When we arrived after a 4½ hour flight we were welcomed by a guard of honour and a delegation of Lebanese came forward to welcome each of us personally. We were given a private car and ours was driven by a very handsome chauffeur who was always smiling and courteous. This was the only time I ever had a motor cycle police escort. The Lebanese people have great charm and warmth and everything was done to make ours a very happy tour.

On our first evening we drove up a long winding road behind Beirut which reminded me of Hong Kong, and at a new hotel Al Bustan there was a reception given by the Minister of Tourism.

The following day we went to Tripoli, a 50 mile drive along a winding road. Driving through the older parts of Beirut we passed the dockyards and the Palestine refugee camps. These wretched people were living in tin shacks held together with string and it was pathetic seeing little children running about in dirty dusty alley ways. There are some half-a-million refugees in the Lebanon waiting for their return to who knows where. In Tripoli we visited the old mosques and houses and we were taken to a large cement plant where we were welcomed in their guest house set in a lovely garden by the sea. There we could swim or go for a boat ride, or relax under gay umbrellas with folk music playing. We took a picnic lunch on tables, a quite sumptuous affair. We were entertained by a belly dancer, an enchanting six year old Bedouin girl who was very accomplished and danced on our tables whilst we lunched.

In the evening we went to the President's Palace for a glittering, highly impressive occasion. The Lebanese women are most elegant in the latest Paris fashions and wore beautiful jewels. Everyone mixed freely and we met many charming Lebanese people. The banquet was delicious but I must say that three main meat courses defeated more than myself. We then retired to the marble terrace overlooking a most beautiful garden and in the warm still night with moonlight sparkling on the sea, we were given coffee and liqueurs.

The following evening we were entertained by the British Ambassador and his wife and were able to talk with many and learn much about the life as seen by Britishers in the Lebanon. We then went to the famous casino and were received on an enormous marble terrace overlooking the sea. It was set in a magnificent situation to capture the glamour of Beirut by night. After being received by the chairman of the casino we had yet another champagne party and were then taken into the casino to dine and see the fabulous show. This had everything including elephants, fine white galloping Arab horses, glamorous girls, a man-made river running through the stage, paddle boots, and even a train and a cavalcade of veteran motor cars. The finale was a raging water fall and girls descending from the ceiling in glittering bubbles, a two hour show which literally flew and left everyone goggle-eyed.

The following day we went to the famous town of Baalbeck, a 55 mile drive into the mountains. On the way we were intrigued to see the Bedouins sprinkled out in isolation on the mountain side, living in tents made of brown sacking, flocks of sheep sharing their accommodation, camels and donkeys tethered close by and children running freely in the most primitive conditions. At Baalbeck we saw the oldest and most famous ruins, the Temple of Jupiter. It was built in the second century A.D. Nearby is the Acropolis with the Temple of Venus still standing. The stones at Baalbeck weigh as much as a thousand tons each and it is quite incredible how the Romans could have constructed them, let alone carried the stones so far from their own land. It makes one wonder how much we have progressed in the last 2,000 years.

On our way home we lunched at the house of a charming Lebanese lady whose sole interest was to promote friendship between the Lebanese and British guests. We were entertained most elegantly to a delicious lunch in her most attractive garden.

At night we went to the Jeita Grotto, ten miles from Beirut. We were whisked in a bubble car across a raging gorge. When we disembarked we were inside the most breathtaking cathedral caves deep in colour, vast silence amongst stalactites millions of years old and woven into a pattern cleverly illuminated. The magnificence and grandeur was a most moving experience. We were given a fashion show by London model girls specially flown out for the occasion.

Later on we were entertained at the Automobile and Yacht Club which is a marina in full view of Beirut and quite magnificent.

On the last day of our trip we paid a visit into Beirut to look at the gold market which was like an Aladdin's Cave, beautiful gold jewellery to be purchased most reasonably. We were welcomed in the shops and could choose at leisure without persuasion or pressure from the shop keepers who treated us to refreshments of ice cold drinks or Turkish coffee.

Our last evening was spent at the home of my company's agent in the Lebanon, a Mr. Mapplebeck, who I had known for many years and was interested to find he had been educated at King William's College in the Isle of Man.

Finally we paid a visit to a Mr. Henri Pharaons' house with magnificent treasures, Chinese porcelain from the fourth century, a marvellous collection of Persian carpets, Byzantine paintings and Bohemian glass. There was time the following day only to say our farewells and as a kindly thought on leaving we were presented with a Cedar of Lebanon tree to plant in our garden. I carried away the most happy memories of a wonderful tour and the desire to return again to the Lebanon.

A fellow guest on this trip was the Rt. Hon. Edward du Cann, M. P., with whom I had long conversations. I felt he was undoubtedly one of our most accomplished businessmen and politicians – a pity he was not holding some important office in government.

I was greatly indebted to Lord Thomson for this opportunity of visiting the Lebanon under such ideal conditions. Unlike Lord Thomson of Fleet who built up his newspaper empire by the aggressive acquisition of one newspaper after another I have explained the different and extraordinary experience I had with companies coming to me and almost begging to be taken over.

In 1971 I became master of a City livery company, the Worshipful Company of Coachmakers and Coach Harness Makers. To those unacquainted with British traditions and customs it might seem a strange anachronism for somebody living in the age of the Concorde and man's conquest of the moon, to be associated with those whose business would appear to be the making of coaches and harness for horses, an obscure and pathetic group of individuals trying to sell wares which nobody any longer wanted.

It was my old friend the late Sir Frederick Handley Page, who revived this city company in the early 1940s after it had lost its old city hall in the bombing of London. He introduced many new members from the aircraft and automobile industries, bringing new blood into this guild.

The keeping alive of some eighty livery companies with their City Halls in which their courts and dinners are held and where every square foot of space is worth a fortune seems incredible. However, these city companies have moved with the times and all are keen to keep alive the old traditions and colourful ceremonies. In July 1973 this city company presented me with their annual award, 'For the founding and development of the Dowty Group'. This was a splendid occasion attended by many representatives from the aircraft industry. No other country can boast these old institutions and the manner in which they are kept alive is proof of the value we place on them.

In 1971 I suffered a grievous loss by the sudden death of our technical director, Tommy Andrews. He had been a tower of strength to our group since he joined us in 1951 and I lost a very close personal friend. He was talking with me in my office one day when he suffered a heart attack. I sent him home and after some difficulty contacted his doctor who to my surprise prescribed walking over the hills! I was shocked when I heard that no cardiograph was taken. This was the second time I had lost a senior director by what I can only describe as medical incompetence.

I remember an amusing incident when Tommy Andrews came to see me on leaving for a holiday. He was a yachtsman and said he was picking up a yacht he had hired in the South of Spain. I casually remarked that I hoped the yacht really existed. Well, as it turned out it didn't and ever since then Tommy thought I was psychic.

In 1972 I was given the Honorary Doctorate of Science from Cranfield Institute of Technology. The Chancellor was my old friend, Lord Kings Norton.

In 1972 the Tewkesbury Chamber of Commerce, with other interests in Tewkesbury and district, sponsored an award fund known as the Sir George Dowty Award, in recognition of the services I had given to promoting industry in the Tewkesbury area. Lord Robens, former head of the National Coal Board, made this announcement and paid tribute to my inventiveness and setting up of the many companies which had brought great prosperity to Tewkesbury and District.

For some time I have been a trustee of the Tewkesbury Museum and of the Abbey Lawn Trust, which is concerned with the development of land and the restoration of the old medieval cottages in the abbey precincts. In 1956 I joined a committee to help the Bishop of Bath and Wells (Jock Henderson), when Bishop of Tewkesbury, to raise considerable funds for the restoration of Tewkesbury Abbey. The appeal was most successful and all the necessary work to the Abbey fabric was completed in four years.

Later that year I again met the Queen when she opened the R.A.F. Museum at Hendon. Air Marshal Sir Dermot Boyle had been the moving spirit behind this imaginative scheme. Sir Arnold Hall, Sir John Lidbury and myself were sponsors of the Sir Sydney Camm Memorial Hall and the Queen, who was accompanied by Lord Carrington, talked with us for some time.

In July, Pershore celebrated the millennium of the building of its Abbey. Princess Alexandra came to the town for that occasion and as president of the Millennium Year activities I presented her with a gift of a silver goblet. She was a most charming person.

I have been associated with many local activities, as chairman of the Worcestershire Association, the Gloucestershire Society, the North Gloucestershire Industrial Savings Committee, President of the

Gloucestershire and South Worcestershire Productivity Association and the Gloucestershire Industrial Safety Group.

In 1972 we set up another Group company, Dowty Leaseholds and Properties, to own our extensive land and buildings. This was done to enable us to defer a substantial tax payment of a million pounds. Credit for this must go to Sir Arthur Norman, Chairman of De La Rue, who alerted me to this possibility. I have found it extraordinary that devices like this can be found notwithstanding our complicated legislation.

I was president of the Old Elizabethans Association (the old boys of Worcester Royal Grammar School). The editor of their journal *The Elizabethan* asked me about relations with my employees. I said that these had always been good and my employees had been kind to me but there was no simple answer to good relationships. These came through a general atmosphere of goodwill on both sides. An employer should strive to make his employees happy by providing good working conditions, recreational facilities, welfare schemes, education services and so on.

From the very early days in business I always made up an employee's wages when he was indisposed and away from work. This was at a time when it was considered unusual treatment. In some cases payments went on for many months.

It was such actions that built up considerable goodwill and I have been amazed at the kindly letters I have received from employees. Even at Christmas time I get cards from employees (anonymous) thanking me for providing them with work and for being a good master.

I hoped that the many managers I employ would deal kindly and sympathetically with employees but occasionally there are those who want to show their importance by taking arbitrary actions.

When asked about my opinion of modern youth I said I was well satisfied with the quality of modern young men but pointed out that things had changed since my early days when my technical education had to be done in the evening after I had finished work. The scientific developments that have taken place in the last fifty years and the resulting expansion of industry have undoubtedly attracted more people and provided them with increasing opportunities.

In 1973 I visited the Paris Air Show at Le Bourget with my son, George. We were two of the few privileged persons to be allowed over the Russian Concorde, the Tupolev TU-144. The Tupolev seemed to me to be a very poor copy of Concorde and I was not impressed with what

I saw. The following day we were at the French Derby when we heard of the disaster that had overtaken this aircraft resulting in its complete destruction and the loss of many lives.

Many of those with long experience in the aircraft industry were greatly concerned when they heard of the terms accepted by Rolls Royce for a contract to build R.B. 211 engines for the Lockheed Company of America. It is well known that any new engine or airframe costs many times the original forecast. This is the pattern of aircraft costs both in this country and abroad. If the government are involved then the taxpayer makes good the cost escalation but when the work is undertaken for an airline the price is fixed and there are penalty clauses for late delivery. When a design introduces new materials then in taking on such a contract Rolls Royce could have had little regard to history. Nevertheless, Mr. Wedgwood Benn as Minister of Technology endorsed these ruinous terms which ultimately led to the government taking over the company. When the conditions of the R.B. 211 contract were known many of us were alarmed and advised our friends to sell Rolls Royce shares. One large company even took out an insurance policy against Rolls Royce becoming insolvent. Yet there were those in the government who expressed surprise when the crash came.

The report of the Rolls Royce failure published by the Ministry of Trade and Industry in July 1973 stated that the Rolls Royce directors accepted a rash commitment which placed at hazard the great Rolls Royce company. But industry was saying this very thing some years before that report was written. The senior directors of Rolls Royce seemed oblivious of the position for their managing director in a letter to me said, 'Our financial difficulties are now behind us,' when two months later the company went into liquidation!

Having been in the aircraft industry all my life I can see the bad image this industry has created by not being realistic in its estimates, and the stupidity in which government contracts are placed.

The design of an airframe or an aero engine is the most exacting of all engineering products, therefore, there is always the need to produce more power or greater strength yet at a minimum weight. Designers are driven to new materials and the use of new techniques. Although laboratory tests may prove satisfactory, weaknesses are invariably only found when the products go into service.

A new design is estimated from drawings on the basis that nothing will go wrong but history has told us that first estimates can be doubled or quadrupled to reach a realistic price. This is due to the many modifications and alterations that are found necessary during development, the time scale slips back and overheads are much greater than those estimated.

But no-one seems to take heed of history. That is why the R.B. 211 and the Concorde, to mention only two recent instances, costs have escalated many times over those first estimated. Of course, if realistic prices were given in the first instance having regard to past experience the costs would be such that no orders would be placed!

Then there is the method by which cost investigation is carried out by government departments. These have no regard to a company's efficiency. As the government allows a percentage profit on cost it pays the manufacturer to be inefficient. The more efficient he is the less profit he makes. Companies should be given a fixed price and if they make a considerable profit that is good for everyone, for even the government takes 52% of the company's profits. This would have many benefits, reduce the number of civil servants and put a premium on efficiency.

Until recently it seemed that China adopted a similar philosophy to that of Emperor Chien Lung, who wrote to George III and said, 'Our celestial empire possesses all things in prolific abundance and lacks no product within its border. We do not need to import the manufactures of outside barbarians in exchange for our own produce.'

But times have changed and China is determined to become a great industrial power capable in time of manufacturing her entire domestic requirements. In the meantime China is doing very considerable business with the U.K. and placed an order worth nearly £20 million with my group for mining equipment.

Nationalised Industries

T HE NATIONALISATION OF industries can cut both ways. If properly organised they can result in economies by the standardisation of equipment and bulk purchasing. On the other hand, as the sole purchaser a nationalised industry can act ruthlessly, stifle development, give no consideration to commercial rights and bring in too many companies, to the point where economic production is not possible.

In 1957 the railways decided, because of wagon damage, to install hydraulic buffers giving greater shock absorbing capacity to replace the old fashioned spring type. The railway sidings on our land at Ashchurch were adapted for tests and the value of hydraulic buffers convincingly demonstrated. A fully laden standard wagon with these buffers was brought harmlessly to rest when travelling at 20 m. p. h. When a similar vehicle was tested but using the standard steel spring buffer the wagon was split from top to bottom, the contents thrown out, the four axle boxes broken into fragments. There was little doubt the hydraulic buffer had arrived.

Large orders were placed and in my report to our shareholders for the year ended March 1957 I was able to say we were producing at a rate of several thousand a week. These buffers had been ordered of a longer length than the conventional English type but similar to continental standards. It had been overlooked that this necessitated a longer chain for coupling wagons. When the first deliveries were put into use it was found that the longer chain dragged on the permanent way! But there was something even more serious. There is a law in this country that the shunter must not go between wagons to connect them, the operation being carried out by means of a pole. The longer chain proved so heavy it was found impossible for even the strongest man to lift the chain. Orders for the buffers were hurriedly cancelled at a heavy cost to the taxpayer.

This was no isolated incident. All goods wagons have no remotely operated wheel brakes. When a train of loaded wagons reaches an incline the train is stopped and the guard walks the whole length of the train applying the brakes on each wagon.

At the bottom of the incline the train is again brought to rest and the guard walks the whole length of the train to place the brakes off! To overcome this inefficient method of operation a production order was placed for remotely operated wheel brakes. Wagons carrying bulk produce, such as coal, are emptied by means of tipping the complete wagon. When the first braked wagon came to the tipping gear it was found that the brakes fouled this apparatus. The production order for brakes was cancelled. Another costly blunder.

The railways thought it would be a good idea to have an automatic wagon coupler for this would save a great deal of time. A committee was set up by the railways to work with my company on this project. To couple a heavily loaded wagon with one that was empty meant coupling wagons standing at different heights, it was also necessary to couple and uncouple wagons when on a very tight curve. Furthermore, these couplings had to couple and uncouple conduits for operating the wheel brakes. This was a complicated exercise and took two or three years to complete, during which time my company spent some £200,000 on this project under the eye of the railway committee.

At last the project was complete, an appointment made with the chairman of the British Transport Commission, the late Lord Robertson (then Sir Brian). A working model of the masterpiece was placed on his table. I waited for his acclaim. But what did he say? 'What a pity we haven't the money.'! To have been kept in the dark over these years, to have allowed private enterprise to continue with this abortive and costly exercise was disgraceful.

But the National Coal Board was no better. We were the first to introduce what is now the worldwide accepted method of supporting the roofs of coal mines by hydraulic supports. A nationalised industry will not accept one company having a monopoly, nevertheless, to bring in too many companies, as the N.C.B. did, was uneconomic and I told them so. There was a period between 1967 and 1971 when mines were being closed and the National Coal Board realised they had too many suppliers. To rationalise the Industry they did not face the issue by ceasing to support companies that had made little or no contribution to research and development. What did they do? They went to the larger suppliers and said, if you want further business from us you must take over the smaller companies and we will then guarantee you a certain volume of business for the next two or three years. In this way the Industry was forced to take-over and pay handsomely for businesses they did not want. The Industrial Reorganisation Corporation (I.R.C.) headed by Mr. Charles Villiers were involved in these manoeuvres with the N.C.B.

As Gullicks had taken over Dobsons of Leicester there were only the smaller companies left. We were urged to acquire the business of Anderson Mavor of Motherwell, the well-known manufacturers of coal cutting machines, but Mr. James Anderson was, at first, against this proposal and we were left to take over Bonser Engineering – a company we did not want for they had little to offer. However Bonser obtained a promise from Lord Robens that if this deal went through our orders from the N.C.B. would be doubled. Six months later N.C.B. officials said they did not know where these extra installations could be used!

Later James Anderson changed his mind and there were many meetings with the I.R.C. and N.C.B. before terms were agreed which made this take-over viable. The N.C.B. promised to place orders for a certain value with Anderson Mavor over the next three years. Both companies took advice from their lawyers and financial people and finally shareholders' meetings took place and the take-over approved. The mechanics were complicated, took a long time and were costly. The deal was to go before the Scottish courts for ratification on a Monday. On the previous Thursday, Anderson Mavor received word from the purchasing department of the N.C.B. that they were withdrawing their commitments for the promised orders on which the take-over of Anderson Mavor had been based – a complete back down on the arrangements agreed with the N.C.B. and the I.R.C. My company received no communication from the N.C.B. or the I.R.C. which was, to say the least, surprising. I tried to speak to Lord Robens but was told he was out of the country and Charles Villiers was too busy to speak to me! Of course, they must both have been highly embarrassed and did not want to speak to me. We had to appear before the Scottish courts with only one working day intervening to say we must withdraw our offer to take over Anderson Mavor.

We had cause to take legal action against the I.R.C. and the N.C.B. to cover our substantial costs and those of Anderson Mavor but how was this possible since the N.C.B. was our only customer? At first they would not accept liability and it was only by exercising considerable pressure from our lawyers we managed to get a contribution towards our costs, but only then on condition that we said nothing about it!

Another objection to the N.C.B. has been their attitude towards commercial rights. When a company undertakes development at its own cost much of what it does never reaches the production stage. So when patents are obtained for successful ideas industry must take into account the outpourings of funds it made on developments which have proved abortive. Nevertheless, the National Coal Board will only use patented products providing they are given demand-free rights permitting competitors to use these patents without paying a cent. The N.C.B. have been known to standardise on inferior designs rather than acknowledge patent rights.

These are some bad features of the nationalised industries. They have unlimited powers and are dictators from whose arbitrary decisions there is no redress.

I have seen many hundreds of millions of pounds squandered by the sheer incompetence of nationalised industries. But their mistakes never come to light – there are no A.G.M.s at which shareholders can question the stewardship of these industrial giants.

It is of interest to reflect on the difference between private enterprise and the nationalised industries. Private enterprise businesses are financed on risk capital which means there must be careful and wise management to ensure profitable operation. The initiative and experience of private enterprise is noticeably lacking in nationalised undertakings.

Under private enterprise failure means liquidation whereas in a nationalised industry this means a greater burden for the tax payer and a government whitewashing of the failure.

We have seen abortive attempts by the government in starting new enterprises such as the ground nut scheme and egg production in Gambia.

Civil servants have a high sense of public duty and are conscientious but their weakness is one of complying and not questioning, to defer and avoid making decisions. The young civil servant must wait for a dead man's shoes whereas those with enterprise can, with private enterprise, find ample opportunity for quick advancement.

Taxation has for long stifled growth of private enterprise. Retained profits have not matched the replacement cost of wasting assets and this has led to weakening of the country's economy.

Government Research Establishments. There is little doubt that a grave waste of public money is spent by the government on the many research establishments up and down the country. It is difficult to find out what these cost the taxpayer but it must run into hundreds of millions of pounds annually, and the work being undertaken, from what I have seen, is of little consequence.

I have visited many of these establishments over the years and find they are hard pressed to get anything useful to do. Certainly industry could not afford to operate research in this way.

The National Research Development Corporation offer all manner of inventions to Industry but complains that Industry takes no interest in their 'golden nuggets'. Out of hundreds of inventions listed few make any sense yet these inventions continue to be listed year after year with no takers.

EDUCATION

IN 1957 I received an invitation from the Master and Fellows of Selwyn College Cambridge, to attend their Shrove Tuesday Feast and to see the university engineering department.

The feast was splendid and held in a hall of the college. I was the only industrialist, the other guests included Lord Adrian, Master of Trinity, Sir Edward Bullock, Director of the Royal College of Music, Sir Humphrey Trevelyan, late the Ambassador in Cairo, the Lord Bishop of Ely, two Members of Parliament, the Master of Haileybury, the Master of Peterhouse, Lord Rothschild and so on, fifteen guests in all. Of the twelve Fellows at the feast, two were lecturers in theology, three in natural sciences, one in zoology, one in history, one in modern languages, one in English, one in soil sciences, one in chemical engineering and one in engineering. The arrangements were first-class and I was shown the greatest hospitality.

On visiting the engineering department I was met by Mr. Welbourn, university lecturer in engineering, who was also Dean of Selwyn College, a very capable man but talkative. It was difficult to get in a question or two. Professor Baker (later Sir John), the head of the engineering department, was an extremely nice man.

However on entering the engineering shops I was shocked to see such great untidiness. Had they been my shops, the manager would have been promptly sacked. Certainly the workshops of the North Gloucestershire Technical College were a model compared to the machine shops at Cambridge.

Men of 20 years of age were engaged in making bench vices, work done in modern secondary schools by boys of 14 or 15. Others were working in the carpentry shop. Even our apprentices do not go into the carpentry shop unless they are trade apprentices. I did not know that men spent their time at university doing these things.

Passing into the laboratories I was surprised at the elementary equipment. In one section there were six sets of equipment for doing simple crippling of flat plates – work I did at night school at the age of 16.

The drawing work being undertaken was similar to that in a secondary technical school by boys of 14. The work I saw at a Cinderford School some months before was every way as good as the drawings being done at Cambridge.

In one of the laboratories I met a Canadian aged 22 determining the out of roundness of a ball bearing using what looked like an inverted egg cup with a compress air jet. He had spent one year on this work but found it so absorbing he was to spend a further year on this investigation!

Another student had spent two years on the flow of liquid in pipes and had an amazing array of equipment for making all sorts of wave patterns which seemed an amusing pastime. When I asked where it was leading I got no constructive answer. Another graduate had spent two years investigating oil leakage past piston rings. When I asked him what he had discovered he made the somewhat astonishing reply that it depended on the accuracy and finish of the rings. When I asked Welbourn why these men spent so long on these particular problems he confided in me that they were not interested in training engineers. Their job was to get men to think mathematically.

He told me that he doubted if half of the men there knew how an internal combustion engine worked. I was horrified.

Even the head of the engineering department seemed to have a bee in his bonnet by designing structures on a basis which he described as the 'Theory of Rusty Hinges'.

I thought I might have seen some advanced training on subjects to make for the balanced engineer. However, it seemed that the products of this university would probably end up goggle-eyed mathematicians with a job in the corner of some stress office. Having spent only a day in the engineering department it may have been presumptuous of me to be so critical. What I saw and heard left me unimpressed.

I was no longer interested in campaigning for Cambridge University graduates but quite satisfied in the products of our local technical colleges. I now understand why, so few university men seem to go places in industry. When I showed the report on this visit to Sir Roy Fedden he said, 'You've seen nothing – go to Oxford!'

I have always been keenly interested in education – particularly in the training of young men for industry. At any time we have some 700 undergoing training in my companies and about 10% are at universities – the red brick ones.

Naturally the most important are those with aptitudes for management. Three times each year I spend a morning with junior executives – those we believe to have management potential. The meeting is informal with the men asking me questions on all manner of subjects ranging from finance to production and selling. From the earliest days in business I have given young men an overall training to help fill the senior positions available within my company.

The most important qualification is practical ability – the ability to manage – to know one's markets – those with technical ability to improve products – but above all to train the all-round businessman. Out of every hundred young men there are perhaps no more than 5% who reach positions of considerable responsibility. Others play less important roles and some are craft apprentices so important to any manufacturing company.

Each year we take on several hundred young men and in their last years they are encouraged to undertake original and practical projects. They have designed and manufactured special equipment for hospitals, for the disabled and have undertaken such jobs as repairing the carillons of Pershore and Tewkesbury Abbeys. They have been responsible for producing specialised equipment to improve productivity in some of our companies.

We have not confined our training to British nationals but have received young men from Ceylon, Egypt, France, Germany, Jamaica, Malta, New Zealand, Saudi Arabia, South America, and Switzerland, and we have sent young engineers to work for the Voluntary Service Overseas in Kenya, Nigeria, the West Indies and Zambia.

Every year my company takes over the Cheltenham Town Hall for our Apprentice Prizegiving, an occasion at which parents and local educationalists attend. Some of our company's products and samples of the apprentice's work are exhibited and an address is given by a guest speaker. Speakers have come from all walks of life, sportsmen, politicians, industrialists, and on one occasion a bishop. There was a time when we invited a well known trade unionist, a Mr. X. His address was quite appalling, the worst we have ever had. Mr. and Mrs. X stayed in my home as my guests. During the evening my wife said from a conversation with Mrs. X it was clear that they were expecting much more than the gift in kind that was usually given. The next morning a pile of notes slipped into Mr. X's hand disappeared in a flash, clearly someone not inexperienced in this art.

Education costs this country £1 per week for every man, woman and child or £3,000 million per year. Much of this huge expenditure is misplaced for many graduates cannot find employment and industrialists complain of decisions such as the constitution of the Robbins Committee which was set up the early sixties. The preamble to the Robbins Report shows little idea of what was needed.

At this time I was greatly concerned at the findings of the Robbins Committee set up to review the pattern of higher education in the light of national needs. It seemed to me to be the most foolish of documents and due no doubt to the constitution of the committee set up by Quintin Hogg. All except one were educationalists and of the eleven committee members, just one came from industry – the chemical industry – and he put in a minority report. The average age of the members was 60.

Through my many interests in Worcester I knew many of their personalities including the dynamic Member of Parliament, Peter Walker. His success in the Heath administration demanded allegiance to the party line for at one time he was an anti-common market man. When I complained to Peter Walker about my feelings of the Robbins Committee he suggested I should meet Quintin Hogg [Lord Hailsham], who was then Secretary of State for Education, and suggested that as I held such strong views on the constitution of the Robbins Committee and their findings that I should have the opportunity of stating these to the minister. A luncheon took place at the Ritz Hotel on 25th March 1964, with the three of us present.

I found the minister to be opinionated and quite rude. He said if industry did not like the way men were to be trained there were plenty of opportunities outside industry. I pointed out that it was by industry we all lived and why had he only selected one industrialist to serve on the committee? He replied that he had thought of putting another industrialist on the committee but could not find one.

About this time a debate on education in the House of Commons was reported in *The Times*. It said that the minister ranged from ranting self confidence to sheer bathos and when the Commons laughed they laughed not with him but at him and when they jeered it was not at the party or the policy but at the man.

In Sir Alec Douglas Home's memoirs he recalls that in the previous October Lord Hailsham had made a complete ass of himself at the Conservative Party conference at Blackpool by his ridiculous antics that had cost him the party leadership. Here is a man without any experience of industry and a fool to wit, deciding what training is best for those entering industry.

I have found educationalists to be ever thus. Any question as to the purpose of education is considered impertinent. The effect of expanding higher education is the over production of graduates and tends to increase the academic qualifications required to enter a career although the ability to pass examinations is not necessarily the best method of finding the practical men industry needs.

Those who complain about the number of technologists and engineers we train compared with other countries omit the numbers in our colleges of advanced technology and similar institutions. If these are included then as a percentage of population Britain trains eight times as many technologists as either Germany or Japan and twice as many as the United States. Industry is more interested in technologists and engineers, the practical people that Industry wants, but university minded people think that colleges of advanced technology should not be mentioned in the same breath. What is clear beyond argument is that young men should be trained for the Industry in which they are eventually to serve. Most industrialists believe young men embarking on an industrial career through universities should spend some time in industry first but there are universities who dislike industry sponsoring young men in this way. There was a Cambridge College that actually described industry as 'cradle snatchers'!

I have been appalled at the absence of cleanliness and tidiness in many university workshop and laboratories, conditions that would not be tolerated in industry. The attitude particularly of some of our older universities is unrealistic and their teaching undertaken by men who have no experience of industry and are out of touch with the realities of life. It is a strange reflection on our educational system that those expected to employ the products of higher education are seldom if ever consulted.

I have been alarmed at the low level of post-graduate work and the way in which tax payers' money is squandered. Men are encouraged to believe that post-graduate work will help them to get better jobs but they should be told they are wasting the most precious years of their lives. Britain has dedicated itself to higher education for a maximum number, an inspiring motive but an expensive one. Perhaps we have not paused to ask the purpose of this. If it is suggested that this is to improve the chances of getting better jobs then men are being misdirected.

There or so called management training schools in profusion, such as staff colleges charging substantial fees. I made the experiment of sending one of my senior men to such a college. I called one mid-morning to see how my man was getting on and found he was playing golf. When he turned up and I asked what work he was doing he said for the previous three weeks he had been studying the life of Wellington.

In 1967 I was given the Honorary Degree of Doctor of Science of Bath University. Receiving the same degree at that time was Sir Eric Ashby, one time Vice Chancellor of Cambridge University and latterly Master of Clare College, Cambridge. Sir Eric published many books and referred on numerous occasions to persons like me, who never went to a university, as amateurs. When I responded on behalf of the graduates I said that I wished the Inland Revenue would accept Sir Eric's view of my status!

The stupidity of Sir Eric's remarks are highlighted by men such as Henry Ford, Lord Nuffield, Edison, Issigonis, Brunel, George Stevenson, Sir Sydney Camm, to mention but a few outstanding engineers, and what about Jesse Boot and Irving Berlin? Edison, the great inventor, passed no examinations. He was untutored and unschooled. In my early days the only illumination I knew was gaslight. Then one day came the electric light – the product of the uneducated Edison. Had this been his only achievement he would still have been one of our greatest benefactors – but he also invented the phonograph and the cine camera.

Lord Kearton has remarked that many of the best engineers have come up 'the hard way' without the benefit of a university training. The men who are all important are the original thinkers – the creative. The late Dr. Bronowski has said that education dulls an inventor's mind and Sir Alec Issigonis that formal education only handicaps the natural designer. The finest examples of architecture were carried out before architects were required to pass examinations and this is true of music, painting and other skills.

A governor of three schools, the Worcester Royal Grammar School, Dean Close Cheltenham and the North Gloucestershire Technical College, I found these meetings boring and so attended infrequently. The number of governors is too many. As a businessman I would not call a directors' meeting to deal with matters of such little importance.

Except maybe for finance I see no reason for taking up the time of so many busy people.

I have no doubt that the soundest feature of our educational system has been the English grammar school. They turn out sound practical young men and I speak with experience for so many of those I employ have reached senior positions and have come up that way.

I do not understand why politicians with little or no experience of our grammar schools should seek to destroy what has proved so excellent.

I have said that I left the Royal Grammar School of Worcester in World War One for economic reasons but I know how much I owe to the education I received which gave me a good start in life. The grammar schools and our local colleges of technology are without their equal in any other country and we should encourage these institutions and not make changes which can prove detrimental.

Sports and Recreation

'WHEN I PLAYED for England at Twickenham' is always a useful ploy when addressing Rugby enthusiasts. There are those who only associate Twickenham with one sport. However in 1962 I played for England at Twickenham against Scotland at curling – that year England won the encounter. Each year the rink of four players on the winning team that wins by the biggest margin receives gold medals, and I was fortunate to be one of the four curlers to achieve that distinction.

The first time I went to winter sports in Switzerland was in 1933 with Mr. A. W. Martyn, sometime my chairman. He took me to Engelberg where he had friends who were curlers but I joined the younger set who were skiers. After this first visit the war years prevented my visiting Switzerland again for many years and it was then that I took up curling. I found Wengen a most agreeable spot, unspoilt and without cars.

The chief curling competitions in Wengen are the Molitor Cup and the Wengen Cup, the former being open to teams from all over Switzerland. In the days when I gave a bit of time to curling I had the satisfaction of winning both these competitions.

For several years I was president of the Wengen Curling Club. Interested in the manner in which a stone 'curled', it seemed to me contrary to what one might expect for if it was subject to the laws of dry friction then the direction a stone takes would be opposite to that we find in practice. As no-one could give me an answer I undertook my own investigation and the conclusions were set out in an article published under my name in the *Scottish Curler* for September 1959. As no-one commented on this theory this undoubtedly explains the phenomenon of the curling stone.

I have always been interested in cricket and was a member of my company's executive team which won the company's championship in 1953. I have been a member of the Worcestershire County Cricket Club since my school days in Worcester, and since coming to Gloucestershire in 1924, a vice-president of the Gloucestershire County Cricket Club.

In the years 1936-1938 a large area of the Park at Arle Court was levelled and a sports pavilion built. The Gloucestershire County Cricket Club was invited to play a match for their beneficiary but this was turned down. The Worcestershire County Cricket Club was invited and readily accepted the invitation, so that for twenty years Worcestershire professionals held a benefit match in Gloucestershire.

Tom Graveney, then Gloucestershire captain, invited me to become chairman of his benefit appeal and a match played at Arle Court provided record receipts for a Gloucestershire professional. The Gloucestershire club was not well served by some of its officers and this resulted in Graveney leaving them.

Living on the borders of Worcestershire and Gloucestershire he threw in his lot with Worcester. There were those who thought I had influenced him to leave Gloucestershire but this was not so. Tom, an international player, had led the county into second place in the championship but had been demoted and the captaincy given to an amateur who had no pretensions to first class cricket. Graveney would not accept this and Worcester were glad to have a player of his abilities.

I felt greatly honoured to be elected President of the Worcestershire County Cricket Club in 1961. At that time the club, like many other county clubs, was in financial difficulties. There seemed no businesslike approach to these problems and for many years there had been a heavy dependence on monies provided by the Supporters Club who were continually being depleted of their funds. I set about putting the club on a sound basis, starting a strong membership drive bringing in thousands of new members, raising subscriptions and so on.

During the five years of my presidency we did not have to draw one penny from the Supporters Club and for the first time for many years became self-supporting. I don't think this change in the club's fortunes was generally appreciated although it gave great personal satisfaction. During my presidency I was helpful in getting Basil d'Oliveira to join the club and to see the club twice win the County Championship in 1964 and 1965.

There were young professional cricketers who, playing during the summer season of six months, had nothing to do during the winter, or at least had menial jobs that would lead them nowhere when their cricketing jobs were over. I took an interest in these young men and tried to help those that were so willing, taking them into one of my companies for training during the off season. In this way I helped some of the young men and one of them, now retired from the game, has a full-time job as an outside representative in one of my companies.

Those were exciting years for me. When I was a schoolboy, Worcester with Northamptonshire were always bottom of the championship table. I hated the visits of Lancashire and Yorkshire, counties with ten times the population of Worcester. The treatment meted out to the Worcestershire bowlers I thought most inhuman! The Lancashire team held Worcestershire in such contempt that when they came for their three days' match they only booked their hotel for two nights. Now all was changed. In winning the 1964 Championship we had a fine team and a bowler with the best England averages, Jim Standen, who had been the West Ham goal keeper when winning the Football Association Cup at Wembley a few months earlier. A unique achievement!

It had taken Worcestershire 65 years in the competition to gain their first championship. The winning of the championship for the second time in 1965 was quite sensational. By mid-July the county had only won two games and by mid-August, even when they had won the next three games, even the sports writers had written them off. In their next game they beat Surrey and then went on to play Hampshire in Bournemouth. I remember arriving from abroad at Heathrow at midday on the last day of the match when my chauffeur told me of their hopeless position. On my way home I heard on the radio of the most sensational happenings with two declarations and finally the Worcester bowlers running through the Hampshire innings in fantastic fashion. Now came their last match against Sussex at Hove. I was there on the second day when victory looked most uncertain. However in the end Worcester were victorious but never before had the championship been won by a team whose only appearance at the top of the table was in the last six minutes of the competition. Shortly before midnight the late Dick Lygon and Joe Lister, the Club Secretary, (now with Yorkshire) arrived at my home showing unmistakeable signs of their ordeal for it could truly be said that no-one could have derived more bliss from such agony.

During my term as president I had the pleasure of entertaining many distinguished visitors on the County Ground including Edward Heath, the Prime Minister.

In 1974, ten years after winning their first County Championship, Worcestershire were to become champions again in circumstances similar to the manner in which they won in 1965. In August they were

30 points behind the leaders but from then on they continued to win their matches although they never were head of the championship table until the last day of the cricket season. I doubt if any team has twice won the championship in such a manner.

I put much time and money into cricket during those years I was President of Worcestershire and one might have thought that those in authority would have had some regard to this. In that period Worcester twice played at Lords in the final of the Gillette Cup. As their president I was refused a box to entertain important visitors and only after heated words was accommodation grudgingly provided. On cricket grounds throughout the country one is treated with courtesy and kindness but not at Lords. Like the Jockey Club the M.C.C. are unfortunately self-perpetuating.

On completing my five years as president of the W.C.C.C. the officers were kind enough to present me with a magnificent Worcestershire Porcelain bowl showing a view of the County Ground.

My wife and I were often guests at Madresfield Court, the Worcestershire seat of the Earl and Countess Beauchamp. They were charming hosts and were great entertainers in their splendid Elizabethan moated home, with wonderful treasures.

A frequent visitor there was the late Sir Walter Monckton when Minister of Labour. He often visited the Worcestershire County Cricket Club when staying at Madresfield. It was he who appointed me to the board of Remploy. As president of the Surrey County Cricket Club he entertained me at the Oval on many occasions and also at the Midland Bank Headquarters in London where for some years he was chairman.

Another person I remember meeting at Madresfield was the local Member of Parliament, the late Sir Gerald Nabarro. He was a man who liked the sound of his own voice, was very opinionated, full of his own importance and quite impossible, his only asset being his attractive wife.

In 1956 I became President of the Cheltenham Town Football Club, the Robins, a position I held for several years. I have been interested in association football from the days when I worked in Hamble and visited the 'Dell' to see Southampton, a team I always followed.

My father had a billiard table, a game I have played all my life. Not a bad player, I won my company's snooker championship in 1956. My biggest break of 66 was on the Rolls Royce Table at Duffield Bank, Derby.

When working with the Gloster Aircraft Company I won their chess championship.

For over thirty years I have bred and raced thoroughbreds. I started due to stupid wartime legislation when no-one was allowed petrol for golf yet owners of racehorses were allowed petrol for racing. I have had good horses although with only a small stud successes came all too infrequently.

My first thoroughbred was a mare called 'The Devil's Lady' and its first race was over hurdles at Cheltenham which it won. It was with great regret that my twin brother, who was responsible for this purchase, died before the race took place. This victory made me think winning races was comparatively easy. I was soon to know better.

It is an interesting hobby but one costing my friends much money. Why they should think every time I run a horse it is sure to win I do not understand!

I was a guest at a Buckingham Palace Garden Party in 1953, a day when I had a horse running in the last race at Bath. Leaving the palace I bought an evening paper but could find no racing results from Bath. The telephone wires to the racecourse had been cut by a gang to prevent away bets reaching the course so that a false price could be obtained on a fancied horse. Ultimately the gang was brought to justice and I am pleased to say my horse won that day.

The only time I was interviewed by the stewards was in connection with the 'Great Metropolitan' of Epsom in 1955 with my horse Babylonian. It had run some days before at Birmingham when my trainer Tom Rimell said that it needed a race. The courses are very different and the horse ridden by different jockeys. I was later told that the enquiry was instituted by the late Lord Rosebery, a senior member of the Jockey Cub, who had confidently expected to win that race himself! The manner in which I was treated at that enquiry was disgraceful and all the more surprising having regard to the form of Lord Rosebery's horses trained by Jack Jarvis, for they had for long been a matter of comment among racegoers.

For some years I have been a member of the Council of the Racehorse Owners Association who have always been for the abolition of bookmakers and to rely entirely on the Totalisator, as is the case in the U.S.A. and France. We are all concerned by bookmakers who 'get at' jockeys and horses. Anti-post betting has always been an invitation to interfere with horses such as that Derby favourite, 'Pontirichio'.

I have never been other than a modest gambler but one day at Worcester in my early days at racing I was asked to place a bet for the stable of £1,000, with a bookmaker known as 'Britain's Largest'. The horse was called 'Comique'. l duly won (I believe at 8/1). The day arrived for payment but nothing was forthcoming and when I approached the bookmaker he had the effrontery to tell me I had placed the money on a horse I had never heard of! I took the matter to Tattersalls – a committee that deals with disputes – and they decided there had been some mistake and reduced the wager by one half!

Many racehorse owners can tell of the malpractices by bookmakers. I once had a good Hyperion colt in a race it should have won, but was surprised to find it out in the betting. During the race I saw this colt pulled from a leading position to back of the field. A good friend of mine who was also connected with a very well known bookmaker, said, 'Don't use G. to ride your horses again, he took £500 out of our books not to be in the first three, but if you tell anyone I have said this I shall have to deny it.'(!) I wrote to my trainers and sent a copy of my letter to Weatherby's saying I would not have that jockey ride for me again but no-one took the trouble to enquire the reasons for my action. The bookmaker who bribed my jockey was the same who diddled me out of my bet on 'Comique'.

It is no good those who say racing is straight – it will never be so long as there are bookmakers. Why don't we get rid of them? I am told on good authority that too many people in high places find it beneficial to keep bookmakers in business.

At one time my horses were trained by one who had been a famous jockey and his stable jockey came from 'down under'. I contributed to his retainer. One day he failed to ride a horse of mine but rode at another meeting for owners outside the stable. When I complained, that jockey never again tried to win on any of my horses and the position became so bad that I wrote to my trainer saying it was obvious my horses were not being well ridden and perhaps I should take them away. My trainer had the same idea and a letter from him crossed with mine suggesting the same action. My horses were sent to Newmarket and started winning at once. In fact one horse won 14 races!

I have won many good races on most English courses including the Greenham Stakes twice. I have made many good friends racing and my daughter married into the Hue-Williams family – Mrs. Vera Hue-Williams being her mother-in-law. Colonel and Mrs. Hue-Williams have won most of the important races, in 1973 the Oaks and in 1974, second in the English Derby and first and second with their two runners in the Irish Derby. I was often their house guest for Goodwood and another guest who was always there was that great character, Lord Carnarvon.

The way owners are taken in by bookmakers and jockeys discourages those who would otherwise take a greater interest in breeding and racing thoroughbreds. Perhaps the only real benefit I obtained from operating my stud was the purchase of land around Cheltenham at a time when the number of my horses was considerable. Later on, when I cut down the size of my stud, the land was in demand for housing and I was able to sell this at a considerable profit.

I have made many good friends racing. The late Jack Olding was quite a character. Then there was my old friend Sir Alan Gordon Smith who entertained so well at Goodwood.

On my regular visits to the Bahamas I have got to know Mr. E.P. Taylor, that great breeder of thoroughbreds, and that amusing character Lord Carnarvon (Porchy) and that splendid steward of the Jockey Club, Lord Abergavenny, a charming man.

I remember with affection the first jockey I had, Sam Wragg. Many jockeys have ridden for me including Sir Gordon Richards and that great horseman Lester Piggott.

I have had numerous trainers, George Beeby, Tom Rimell, Fred Rimell, Gerry Wilson, Frenchie Nicholson, Mrs. Johnson Houghton, Sir Gordon Richards, Sam Armstrong, Farnham Maxwell, Peter Nelson, Boyd, (Scotland) and Bill Marshall.

Until a few years ago I played tennis but I have now given this up in favour of golf. I still enjoy swimming, bridge, curling and snooker so I have interests in many sports.

During the late thirties I collected 'British Occupational' stamps. Under the strain of war I found this hobby incompatible with good health for I needed a different form of relaxation. To my later regret I sold my stamps at Harmers' in a two-day sale.

To compensate for this I later became interested in antique silver and made an important collection of the work of Paul Lamerie. These pieces were acquired at a time when they were not selling at the high prices they are now.

For several years my friend Sir Stanley Rouse invited me and my wife as his guests to the F. A. Cup Final at Wembley and we were also the guests of Bill Fallowfield, secretary of the Rugby League, at the Wembley Final.

The late Sir Godfrey Ince and the late Sir Adrian (Jimmy) Jarvis were members of Wimbledon and entertained me there on numerous occasions.

Another sport I indulged in for a short time was water skiing in the South of France and the Bahamas.

I have never cared for some sports such as greyhound racing but many years ago I was invited by the directors of the Cheltenham and Gloucester Greyhound Track to be their guest on their opening night. Without knowledge of the racing ability of the dogs I had my card marked by one of the directors. Strange to say I won on each of the first six races! At that point I quit and have never attended a greyhound track since.

Speaking of gambling luck, I had a somewhat similar experience in Malta. I was invited by my manager to look at their new casino. Casinos have never had any interest for me but I accepted this invitation and having looked at the various games of chance I decided to leave, upon which my manager said, 'Are you not going to try your luck?' Taking a £5 note from my pocket I obtained some chips and went to a roulette table and put the equivalent of 5s. on number 7 which promptly turned up. I took my winnings and put another 5s. bet on number 17 which also turned up. I took my winnings and put a 5s. bet on number 27 and strange to say that number turned up. The chances of betting on three numbers which all turn up in succession must be millions to one. At that point I decided I had had enough and left.

I must admit to having never enjoyed those occasions when invited by my friends for a weekend's yachting. I remember days with Donald Campbell and others but the one that remains firmly in my memory was with 'Dopey' Lingham when managing director of Heston Aircraft. As a novice I was made cook. It was dreadful. Everything I cooked tasted of paraffin. It rained continually and I was wet and miserable. The bunk was so uncomfortable I could not sleep. Wet, filthy, starved and sleepless I had to undergo the final mockery of pretending it was great fun.

Some people have to earn their living on the sea. For them I am deeply sorry but it is sheer madness that makes yachtsmen believe they are doing it for pleasure, refusing to admit that yachting is plain watery hell!

Throughout our company's history we have supported recreational facilities for our employees. In 1936, when we had little funds, we gave £250 to the setting up of a sports club. The following year we spent £400 levelling the park at Arle Court for a sports ground. Since then we have spent lavishly on providing and equipping our various companies with six sports pavilions and sports grounds – skittle alleys – a first class bowling green – and we cater for almost every form of sport.

On my many visits to the U.S.A. I have been taken to see American football by my friend Edgar Cullman, president of General Cigar.

I found American football quite a revelation. Each side brings enough players for three teams – while 11 are playing 22 rest on the sidelines. The players are undoubtedly delicate for I saw the teams changed twice in the first few minutes. A few short bursts of energy obviously places too great a strain on the players.

There is 'time out' for refreshments and the players frequently go into conference. When this is over they trot back to the scene of play.

To make matters confusing there is a large clock with hands moving anti-clockwise. These seem to stop and go without reference to the proceedings.

The highlight of the afternoon is a parade of girls with bands and banners. That is something worth watching.

In 1963 on a visit to America I spent some time with Ed and Dodie Walton, old friends of ours, at a fine hotel in the White Mountains of New Hampshire. The journey was quite interesting for I flew from New York to Albany and then picked up a woman bush pilot who had a small single engine light plane and flew us to the mountain resort of Mountain View. It was an interesting flight with the plane piled with suitcases around us. The outdoor activities there were bowling and golf and it was here I started to take my first lessons in golf from the resident professional. On leaving this hotel the same lady pilot flew us to Montreal. This was again an interesting flight and approaching the airport in our small plane we seemed to be surrounded by many airliners. It was extraordinary but this was the only means of getting back to civilisation; there was no other way.

Family Life and General

I was over fifty years of age before my first child was born. We spent many family holidays together in the South of France and in Switzerland. Both my children are accomplished skiers.

My daughter was educated at Westonbirt, Gloucestershire, Chatelard in Switzerland and at Marlborough when that college became co-educational. Finally she went to Oxford University to read Japanese.

Her coming out party took place at Brockhampton Park – a beautiful setting – which included a splendid fun fair and there were many distinguished guests. She married Peter Lilley at St. Clements Danes, London, in December 1971. The Bishop of Gloucester officiated.

My son was educated at Dean Close Junior School, Cheltenham, where he did well both at his studies and sport, and was Captain of cricket. I always tried to watch him play whenever I could and was particularly delighted over one game against Cheltenham College when his side had lost several cheap wickets and he and his partner hit off the 60 runs to win. My son went on to Marlborough where he was later joined by his sister.

At an early age when at Marlborough he elected certain subjects for his A levels saying he did not want to become an engineer. Later he changed his mind so although he passed four A levels he had taken neither mathematics or physics. It was decided he should take these at the North Gloucestershire College of Technology. He worked exceedingly hard from early morning to late at night and passed both these subjects easily.

Although a keen golfer he would not take time off from his studies so trying to induce him to get exercise I promised him a car if he would do a round of golf at not more than ten strokes over par. This he did in two months!

He visited several universities and decided to go to Brunel in the September. In July I mentioned to him another possibility, going straight into a new business I had started outside the Dowty Group with the object of taking control. He was already aware of the products produced by this company. Having discussed this proposition with others he made the decision to go straight into industry. I did not influence him in any way but applauded his decision.

In 1974 he became a Liveryman by Patrimony of the Worshipful Company of Coachmakers and Coach Harness Makers. This took place in the City Hall of the Barber Surgeons.

For many years we all took holidays together, skiing in Switzerland in the winter and at Val d'Esquiere during the summer.

I had several black cats at my Cheltenham headquarters. One in particular had a habit of hiding in my desk and then suddenly emerging and crawling over my papers in the middle of a discussion – it could be disconcerting for my visitors but hilarious for me.

When I was single the cats always came to my bedroom in the morning. When I was married and brought my wife back after our honeymoon, the first morning of our return one of the cats came into the bedroom, jumped onto the bed between us, dug his claws in and growled at my wife. When ejected from the room he found his way back by another window, walking along a balcony, and repeated his performance. How possessive can animals be!

Indeed I've been a cat lover since my early boyhood. All my cats at Arle Court were great characters. One was an accomplished fisherman who sat tirelessly by the lake. Having mesmerised the fish he pawed them out and brought them to me for admiration.

Again, a bird flew into the living room and could not find its way out. It fluttered around the room seeking an escape. One of my cats in the room rolled over onto its back and stared at the unfortunate bird flying overhead, apparently mesmerising it for the bird dropped straight into the cat's mouth!

A particular favourite was a long-haired black cat with a white front and blue eyes. This was the one with the engaging propensity for office-living. It would sit for hours in the waste paper basket or in an open drawer, waiting patiently for me to finish my work. In the evening all three cats would line up at my door, asking to be taken for a walk. Then they ran behind me as I strolled round the gardens, just as dogs would have done. They all lived until they were over 20 years of age and are buried in a plot of land on the estate.

I have referred to my long association with the Isle of Man. When I came to reside in Cheltenham in 1924 I found in the parish church the longest memorial I have ever seen, running to some 600 words, extolling the virtues of a Captain Henry Skillicorn, a native of that island. In 1738 it was he who made Cheltenham a spa by setting up buildings over the springs and planting avenues of trees. He made the town famous for its curative waters which were patronised by royalty, and Cheltenham developed from a village to a town of considerable size.

It was 200 years later when the town had become renowned as the home for retired colonels from the Colonial Service, when the Promenade was full of bath chairs, and a joke on our music halls, that I started my business which grew rapidly and encouraged others to come to this delightful part of England to set up industries.

In 1969 I was giving thought to passing on my responsibilities in the day to day overseeing of the operations of my many companies. By moving to the Isle of Man I would be sufficiently away as not to be tempted to oversee this day to day work as I had always done, but close enough – one hour by executive plane – to make for easy access.

At that time I purchased Silverburn Farm near Castletown, Isle of Man. I did this in order to take possession of Cly-ny-Mona, a house with 20 acres of land situated north of the farm, but which the owners would not sell separately, Soon afterwards I sold off the farm as I did not want to retain this.

As a house in which to retire, this property was too small but it gave me the opportunity of looking round so that in 1972 I found a Queen Anne house near Castletown, standing in 20 acres of ground. This house was built in 1714 and has its original panelling, staircase and floors, with excellent views.

My daughter and her husband, who had also moved over to the Isle of Man some months earlier, took over Cly-ny-Mona when I moved into the Balladoole house.

At the beginning of the war the Manx regiment was stationed in Gloucestershire a few miles from my headquarters, and it gave me great pleasure to entertain them at that time. Later on when moved overseas, they became prisoners of war for several years and I lost touch with them. When I came to settle in the Isle of Man in 1973 I was surprised to be welcomed by a Mr. Dick Gawne, an officer of that Manx regiment who I had entertained in Cheltenham. He is now Captain of the parish of Rushen and has a fine estate about two miles away from mine.

LATER LIFE

R EMAINING CHAIRMAN AND Chief Executive of my company kept me busy yet relieved me of the daily chores and the many calls on my time for social, charitable and educational activities in which I had become involved. I resigned from these outside interests which in later years had taken much of my time.

The Queen's Award to Industry is sought after for it signifies those companies that are deemed worthy of recognition. My companies have won this award on many occasions both for technical innovation and export achievements, indicating our ability to develop important new products and then to substantially increase our exports. Few companies get these awards on both counts simultaneously.

I frequently found myself in the company of Canadians having married one. Such was Ted Leather (now Sir Edwin), Governor of Bermuda. He lived in Batheaston, near Bath, having been Conservative M.P. for many years of a Somerset constituency. He gave distinguished service to the Conservative cause.

Ted was nothing if not a showman. He greeted us one Sunday in a decorative stole having come from a local church where he had been preaching. One Christmas he put on all the old English customs including the suckling pig complete with apple, which he paraded in front of his guests.

I have always taken considerable interest in advertisements for I felt no advertisement was worthwhile unless it was arresting. A mere glance is about all an advertisement ever gets and it should, therefore, be able to tell its story at a glance. The number of words should be an absolute minimum. It is surprising how difficult it has been to get my companies to accept this. Left to their own devices they would fill advertisements with written matter which no-one would read.

Engineers develop a private patter of their own and do not realise that no-one is desperately anxious to read their words. A technical advertisement must combine sound technical interest with compelling eye appeal. One advertisement correctly written and practically presented, published in the right media, can contact more people in one morning than a salesman can meet in a lifetime.

Our notepaper, catalogues, advertisements and transport vehicles use the name 'Dowty' with a distinctive and original typeface. Although it has a modern look it was first used for a catalogue prepared for the Paris Air Show in 1935.

I like spending some of my time in quiet contemplation but I could never understand those who spend all their life in this way in a monastery. Nevertheless I was interested in monastic life and twice visited Prinknash Abbey where I was kindly welcomed by the Abbot. The quiet atmosphere so removed from the day to day hustle of the world and the experience of complete silence, especially during meals, was a refreshing experience.

I find the attitude of some Members of Parliament who take little or no interest in their constituencies quite extraordinary.

An example of a first class member who takes the greatest interest in the people he represents is Peter Walker of Worcester. On the other there was a member representing Cheltenham (now deceased) who never came near one of my factories – refused every invitation I sent him, except on one occasion, the 25th birthday of the founding of my company, when we held a large cocktail party. He came in at one door,

Sir George and family with Dowty Hawker Siddeley Dove

moved swiftly towards the exit and was gone in thirty seconds! As the largest employer in the area he never once wrote, telephoned or had a conversation with me. I can never understand an M.P. who takes no interest in the key developments in his constituency.

These days I live on the Isle of Man, 40 minutes away by plane from my Cheltenham headquarters.

At my age – I'm now approaching my mid-seventies – I still work a full week, keeping au fait with the affairs of the Group just as I have always done. I've never used any city or town club as I don't believe in them. They are generally a waste of time.

At my home near Castletown my secretary arrives at 9 a.m. By that time I have had my bath, breakfasted and read the English newspapers. I remain in my office till 6.30 p.m., going downstairs for a drink before dinner and to check whether there's anything worthwhile on television. Usually there isn't. So unless I am entertaining guests or going out to friends I go back to my office for another session. It's a great thing to lead a full life. I'm never bored.

So at my age the pattern of my working week is as planned and regulated as it has always been. I go through the designs and drawings that people send me, read all the minutes of the board meetings of our 20 or so subsidiary companies, and study the quarterly balance sheets. I correspond regularly and at a personal level with the managing directors of overseas companies. I read the English and American technical magazines and always keep a razor-blade within reach to cut out anything of interest. I read all advertisements and find they are often of more interest than technical articles.

Elsewhere in this book I trace the growth of my business empire; the way it was achieved, the financial hazards that confronted me on my single-minded drive towards success. I have never remotely looked on myself as materialistic. My sister said the other day, 'Status and success have come so gradually to George that I don't think he even considers them. I don't think he realises what he has achieved.'

I had a tendency to write letters home in verse. In many I made references to the eccentricities of my various landladies. For instance there was one who terribly anti-catholic – she was a Wesleyan clergyman's widow. I chuckled to myself as I led her to believe I had catholic sympathies and told my mother all about this in light-hearted rhyming verse.

I made over 200 patents which have been the backbone of my company's success.

I have always lived modestly. I don't hunt, shoot, fish or own a yacht. All my life I have worked hard and have never had time to indulge in such pursuits. Many of my friends in Gloucestershire, Worcestershire and on the Isle of Man contemplate my possessions and decide I am wealthy. Compared with many I am. I have a pleasant estate and a good staff.

The Common Market referendum is now a distant memory. I was against entry and supported that cause financially

I am not a political animal but once I gave a gift to the Tory party. It was my protest over the proposals to nationalise the aircraft industry.

Bob Hunt, my deputy chairman responsible for the aviation side, joined me 40 or so years ago as an apprentice. He wrote to me and asked for an interview. He'd just left the grammar school in Cheltenham and wanted to study engineering at the North Gloucestershire Technical College. Bob made quite a bold impression. 'I'd like a job, Mr. Dowty, with you but I'd also like some time off – at least a day a week – to attend the technical college.'

I hadn't come up against this kind of condition before, but I liked his approach and sensed his potential ability. 'All right,' I said. 'We'll strike a bargain. I'll take you on, providing you're prepared to put in three evenings a week and Saturday mornings.' I didn't consider it unreasonable at the time. He started with me in May 1935 at 7s. 6d. a week. I also told him I couldn't guarantee any continuity of employment because I wasn't sure about the future of the company. But he has been with me for 40 years.

I have always believed implicitly in the policy of promoting where possible from within the company. A number of the current senior directors and subsidiary directors have progressed all the way from apprenticeship with us.

Anyone in industry will confirm the sense of loyalty in the Dowty Group. That is one reason why, relatively speaking, we have few union problems. In the early days when we were working late, I made sure I had the flasks and sandwiches to keep everyone happy.

If an employee had a personal worry I often heard about it and tried to help. Certainly until the companies expanded and the work-force increased I knew most of the men by their Christian name. There was one employee who joined me first in 1934 and later became my chief engineer at Dowty Mining, Reg Elmes. He had a minor domestic anxiety and I arranged, indirectly, for the family to have a holiday.

'One thing about George, he has never forgotten the people who helped him in the difficult days,' was what I heard told the other day. That's the kind of unsolicited compliment one hears.

I always liked to surround myself with men whose work I know and trust. Usually my judgement is sound. Dennis Bridges, who retired as my chief designer at Dowty Rotol after 40 years, came to Gloster Aircraft Company as a 15-year-old office boy. He stayed on for seven years to do his apprenticeship and in the meantime I left.

He probably thought I had forgotten even his name. But one day he wrote to me for a job. I didn't even bother to interview him. I knew his value and he was soon on my pay-roll.

I am receptive to the ideas of the young. Influenced by my own frustrating start in business and the way my spirit was so nearly broken by the mindless and arbitrary rejection of my ideas, I have stretched out a hand of encouragement to young employees who have something worth saying and offering.

For instance, one of my employees, Hillier, invented printed gaskets but his superior, Bestow, wanted no part of it, but I saw merit in the product and took it up.

Three times a year I have a question-and-answer session with a group of my young 'graduates'. They are told to prepare questions to put to me. It gives me a wonderful chance of assessing the quality of our up and coming young men. Before I meet them I study a portfolio, containing each young man's background and photograph, so when I answer their questions I know who I am talking to. The positive thinking of these young people fills me with optimism.

I have a fetish for tidiness. My desk is never cluttered. Even the telephones and the ashtrays are out of sight. I made sure of that – I designed them myself.

I don't think anyone would indict me as a hypocrite. I have always said exactly what I mean without ambiguity. Perhaps I should now and then have been a little more tactful but I am not adept at cloaking my true feelings.

I give a man promotion because he deserves it and not for his membership of a secret society or his religious or political persuasion.

What other qualities have I? Arguably I'm a bluffer – certainly in the early days. The bluff was based on a measure of courage, hope and determination to succeed.

I have never told anybody something could not be done. Whatever

Dowty Headquarters at Arle Court with offices and factories

Dowty Rotol plant at Staverton

the problems, I turned round and made jolly well sure it could. I accepted the order first – and coped with the problems afterwards. Looking back, I know it was justified, This was a vital part of my philosophy. And it was a part that the late Lord Beaverbrook admired.

As a young man I spent many of my evenings thumbing my way through the technical books and magazines. My sister Jess tells the story of when I was 13 or 14 and went on a picnic I was lying under a hedge reading a book when one of the girls crept up behind me and snatched the book away. She thought it was some blood and thunder. In fact, the book was 'Integral Calculus'. I have never stopped reading and learning. I am now the head of a large business with extends across the world. But I continue, after a full day in the office, to ponder more ideas.

Throughout my life, status has never been important to me. Colleagues remind me that I had to be persuaded in the 1950s to buy a Rolls Royce. It was Sir Roy Fedden who put the pressure on me. 'You can't drive up as President of the Society of British Aircraft Constructors without a Rolls. It isn't the thing for a British industrialist to arrive in a Cadillac and so soon after a knighthood. '

The following passage is largely a repeat of p. 85 [I had received a letter from the Prime Minister on my birthday, April 27th 1956, saying that, subject to my agreement, he would recommend Her Majesty to confer on me the honour of knighthood. Quite a birthday present.

This was something I had never sought. But I was delighted to think that my work had been considered worthy of this recognition.

Having built up my business from nothing I had brought considerable employment to the West of England. There was also the substantial contribution I had made to the war effort. I was president of the Royal Aeronautical Society in coronation year when the Queen was patron of that society. I had been chairman of the North Gloucestershire Disablement Committee for some time and a director of Remploy.

I attended Buckingham Palace on July 12th, accompanied by my wife and sister, to receive the accolade.

The ceremony was impressive. Among those knighted at the same time were my friends Sir Alfred Pugsley, Sir Leonard Hutton and Sir Graham Rowlandson. Others included Sir Solly (now Lord) Zuckerman and Sir Eric Ashby, who had once referred to me as an amateur because I hadn't been to a university!]

As I have frequently asserted, the whole of our country's economy is based on industry, primarily those who produce for export. But looking at the manner in which honours are given one would never suspect this.

In the Birthday Honours of May 1956, 32 Knight Bachelors were created. Yet none was given for services to industry, save remotely one for services to shipping. Even my knighthood was for services to the disabled.

Knighthoods seem mainly for political and public services, to civil servants and sportsmen. This is due, no doubt, to honours being recommended to the Queen by the Prime Minister and with advice from senior civil servants.

The following part is in handwriting:

Higher education is fabulously costly and is largely undertaken in the belief that it is necessary for the masses.

University degrees are now as common as trouser buttons and often are nothing like as useful.

The great men of this world have been those born with natural abilities – and those who in most cases have never been to a university or yet passed an examination.

But most nations are being led astray by educationalists for the sole purpose of gratifying their own ambitions.

POSTSCRIPT

No further manuscript has been found and Sir George continued to work until his death. He died at his home in the Isle of Man on 2nd December 1975 surrounded by his family. Sir George had been treated for cancer in his final years. Tragically during the course of treatment his physician Gordon Hamilton Fairley, Professor of Medical Oncology at St.Bartholomew's Hospital in London, was killed by an IRA terrorist bomb in October 1975.

INDEX

This index includes personal and company names, places and aircraft.

Abergavenny, Lord 139
Ades, Remy 112
Adrian, Lord 126
Ajax, Canada 77, 112
Ajax International 77, 112
Albany, USA 141
Alberta, Canada 55, 90
Aldershot, Hants 21–23
Aldoma, Canada 91
Alexandra, Princess 117
Alexandra Palace, London 46
Allen, Bob 60, 99
Alvis Stalwart, boat 98
America, American 29, 46–9, 55, 58-61, 68-9, 72, 77-8, 82, 87, 94, 98-100, 112, 119, 128, 141, 147; *see also* United States
Amery, Julian 100-1
Anderson, James 123
Anderson Mavor 123
Andoversford, Glos 80
Andrews, Tommy (T D H) 80, 116-17
Anson, aircraft 54, 56, 70
Arabia 128
Arle Court, Cheltenham 39-40, 44-5, 51–53, 62-3, 81, 96, 109, 134, 141, 143, 150
Arlette 39
Arlex 72
Armstrong, Sam 139
Armstrong Whitworth 44
Ashby, Sir Eric 85, 131, 151
Ashchurch, Glos 52, 63, 70, 80, 87, 92, 105, 121
Atikokan, Canada 91
Atkins, Ivor 16
Atworth, Wilts 75
Australia 23, 33, 79, 92
Austria 102
Auto-gyro, Cierva, aircraft 21, 22, 26
Aviquipo 48, 49, 58
Avon (Stratford), River 13, 14

Avro aircraft 19, 21

Baalbeck, Lebanon 114-15
Babisson, J T C Moore 62
Bahamas 78, 112, 139, 140
Baker, Sir John 126
Balladoole, Isle of Man 144
Banbury, Mr 84
Banff, Canada 55
Bantam, aircraft 19
Barber, Martin Fountain 52, 63, 76
Barber, Lionel 84, 93
Barracuda, aircraft 56
Basilisk, aircraft 19
Bastow, Frank 22
Bat, aircraft 19
Bata, Tom 54
Batawa, Canada 54
Bath, Som 131, 137
Batheaston, Som 145
Beauchamp, Earl and Countess 136
Beaufort, aircraft 56
Beaufort, Duke 56, 105, 113
Beaverbrook, Lord 55, 56, 63, 96, 151
Beeby, George 139
Beharrell, G E 37
Beirut, Lebanon 113–15
Belgium 44, 92
Belliss and Morcom Ltd 19
Bendix, company 29
Bendy, company 30
Benn, Anthony Wedgwood 119
Berkeley Power Station, Glos 80
Berlin, Germany 131
Bermuda 54, 145
Bickell, Jack 56
Birmingham 15, 20, 25, 53, 63, 137
Bison, aircraft 22
Bjarne Sjong, company37
Blackburn, aircraft company 44, 111
Blackpool, Lancs 70, 130

Blazdell, Charles 22
Blenheim, aircraft 56
Bleriot, Louis 16
Boeing, company 49
Bolingbroke, aircraft 54
Bombay (Mumbai), India 106
Bonn, Germany 92
Bonser Engineering 123
Bookham Surrey 57
Borg Warner 99
Boston, aircraft 47, 60
Botha, aircraft 56
Bottwood, Canada 69
Boulton Paul Aircraft, company 19, 44,
    62, 74, 105-6, 112
Bourget, Le, France 118
Bournemouth 135
Bowstead, Joe 30, 33, 81
Boyd, George 139
Boyle, Sir Dermot 117
Braille micrometers 81
Bramson, M L 28
Brauer, Herr 36, 37, 72
Breckinridge, Henry 59
Breda, company 36
Bredon Hill, Worcs 13, 14
Breeden, Wilmot, company 40
Breguet design 27
Brickell family 16
Bristol 38, 83
Bristol Aeroplace Co 38, 44, 64, 66, 81,
    96-7, 100
Britannia, aircraft 96
Britannic, ship 55
British Aerial Transport 19, 20, 69
British Aircraft, company 44, 75, 103, 151
British Airways 44
British Columbia, Canada 90
British Transport Commission 122
Britten, Benjamin 102
Brockhampton Park, Herefs 80, 142
Brockworth, Glos 25, 51
Bronowski, Jacob 131
Brunel, Isambard 131
Brunel University, 142
Buckingham Palace, London 84, 85, 137,
    151
Buehler, John 98
Buffalo, USA 59
Bulldog, aircraft 38
Buna rubber 36
Burroughes, Hugh 24
Busk. Edward 46

Cairo, Egypt 126

California, USA 25, 27, 49
Cambridge 46, 89, 126, 127, 130, 131
Camm, Sir Sydney 42, 62, 69, 86, 109,
    117, 131
Campbell, Donald 98, 140
Canada, Canadian 19, 37, 44, 50, 54, 54-6,
    59-60, 69, 75, 77, 79, 80, 83, 89-92,
    112, 127, 145
Cann, Edward du 116
Carey, Fred 79
Carnarvon. Lord 139
Carrington, Lord 104, 117
Carter, Tubby 17
Carter, W G 62
Castletown, Isle of Man 41, 144, 147
Cessna, aircraft 49, 52, 53
Ceylon 128
Chadwick, Roy 22, 35, 62, 74
Charingworth, Glos 14
Charlton Kings, Glos 39
Chatelard, Switzerland 142
Cheltenham. Glos 24–26, 32-3, 38-9, 41,
    43, 45, 52, 58, 62-3, 65, 70, 72, 74,
    84-5, 89, 105, 109, 113, 128, 131, 136-7,
    139-40, 142-44, 146-48
Chicago, USA 99
China, Chinese 107, 115, 120
Chipperfield's Circus 13
Cierva, aircraft 21, 22, 30, 37, 44
Cinderford, Glos 127
Cirencester, Glos 52, 63, 87
Clarkson, Christopher 82
Cleeve Hill, Glos 72
Cleveland, USA 49, 78
Clore, Sir Charles 57
Cobham, Sir Alan 33
Cockshot Plough, company 54
Coghlin, B J, company 37, 50
Colwell, Arch 78
Comper, aircraft company 29, 36
Concorde, aircraft 107, 108, 116, 118, 120
Coniston, Lake (Water) 98
Constance, Lake 36
Constantinesco Interrupter Gear 17
Conway, H G 62
Cook, Thomas 20
Corsham, Wilts 64–66
Courtaulds, company 52
Coventry, Warwicks 31, 53, 58
Coventry Precision 79, 112
Cowes, Wight 35
Cranfield Institute 117
Crawford, Fred 78
Cray C G 33
Cricklewood, London 24

Cripps, Sir Stafford 66, 67
Cropthorne, Worcs 14
Cullman, Edgar 141
Currie, James 46
Czechoslovakia 92

Darby, Ormonde 37
Davis, Stuart 74
Davis Wynn and Andrews 80
Dawson City, Canada 89
Dayton, USA 60, 99
De Havilland aircraft company 27, 36, 40,
    44, 75, 78
De La Rue, company 118
Dean Close, Cheltenham 82, 131, 142
Decca, company 45
Denmark 36-7, 44
Devonshire 14
Dexter, John 33, 81
Dobson, Sir Roy 74
Dobsons, company 123
Doi, Mr 34
d'Oliveira, Basil 134
Dominee, aircraft 56
Dorman, Sir Maurice 101
Douglas, aircraft company 27, 49, 68
Dowty, Edward Flexton (twin brother)13,
    16-18, 21, 30, 70-1, 82, 137
Dowty, George (Revd) 14
Dowty, John 14
Dowty, John Goddard 14
Dowty, Joseph 14, 22
Dowty, Laura 13
Dowty, Robert 58, 104
Dowty, Virginia 80
Dowty, William 13
Dowty group of companies, passim
Duffield Bank, Derby136
Duke, Neville 104
Duncan, Barney 22, 80, 81
Dunlop, company 20, 27, 31, 32, 37, 63,
    111
Dunrossil, Lord 85, 87
Dusseldorf, Germany 92

Eccles, Lord 88
Eckner, Herr 36
Edison, Thomas 131
Edmonton, Canada 89, 90
Egypt 128
Elektronmetall, company 36
Elgar, Sir Edward 16
Elliott, R B 19
Elmes, Reg 148
Elmley Castle, Worcs 14

Ely, Bishop of 126
Engelberg , Switzerland
English Electric, aircraft company 96
Epsom racecourse, Surrey 137
Errol, Lord (Freddie) 89
Evans, Stanley 25, 66
Evans-Hemming, Mr 53
Evesham, Worcs 14, 15, 107
Evetts, Sir John 97

Fairchild, aircraft 47
Fairey, aircraft company 26, 79
Fairley, Gordon Hamilton 152
Falconbridge, Canada 91
Fallowfield, Bill 139
Farnborough, Hants 27, 78, 83, 104
Faudi, Herr und Frau 36
Fedden, Sir Roy 55, 62, 69–71, 86-7, 96,
    98, 127, 151
Fell, Jesse 39, 84, 85
Fell, Sidney 15
Feltham London 35
Finland 37-8
Fisher, Tom 14
Fladbury, Worcs 13
Flamingo, aircraft 56
Fokker, aircraft company 37, 41
Folland, aircraft company 24–27, 37, 46,
    62, 83, 96
Forbes, Sir Archibald 64
Ford, Henry 131
Ford Brothers, company 59
Fountain-Barber, see Barber
France 30, 55, 70, 87, 92, 128, 137, 140,
    142
Frankfurt, Germany 36
Freeman, Sir Wilfrid 68
Frost, Revd 18
Fuji, Mount 107
Fulmar, aircraft 56

Gambia 124
Gamecock, aircraft 25
Gannet, aircraft 79
Garvey, Sir Ronald 106
Gauntlet, aircraft 25, 27, 37-8
Gawne, Dick 144
Germany, German 36, 37, 41, 43-4, 51, 70,
    72, 87, 92, 96, 100, 104, 128, 130
Gibb, Sir Claude 89
Gifu, Japan 107
Gillette Stephens, company 52, 57
Gloster, aircraft company 21, 25, 27, 30–
    33, 36–41, 44, 46, 51, 56, 60, 62, 79,
    83, 96, 97, 102, 112, 136, 149

Gloucester 4, 25, 71, 109, 113, 140, 142
Gloucestershire 7, 24, 25, 81, 85, 105, 111, 117, 118, 126, 131, 133, 134, 142, 144, 148, 151
Gnat, aircraft 83, 96
Goodwood, Sussex 139
Graf Zeppelin 36
Granville, Lord (Edgar) 45
Graveney, Tom 134
Grayswood Hill, Surrey 75
Greer, Mr 59
Grindley automatics 65
Guilmant, Alexander 16
Gull, aircraft 36
Gullicks, company 123
Gurney, Winterbotham and 29
Gwynne. Mr 66

Haileybury College, Herts 126
Hailsham, Lord 75, 129
Halford, Major 62, 78
Halifax, Canada 55
Halifax, aircraft 56, 64, 96
Hall, Sir Arnold 117
Hamble, Hants 22–24, 46, 74, 77, 80, 83, 136
Hamel, Gustav 16
Hamilton, Gordon 152
Hamilton, Sir William 98
Hampden, aircraft 56
New Hampshire, USA 135, 141
Handley Page, aircraft company 24, 44, 64, 79, 96, 116
Handley-Page, Sir Frederick 96
Harley, Sir Harry 79
Harmers, philatelists 139
Harrier, aircraft 109
Harrow, London 72
Hart, aircraft 56
Haslemere, Surrey 75
Havilland, De, aircraft company 27, 36, 40, 44, 75, 78
Hawk, aircraft 36
Hawker, Hawker Siddeley, aircraft company 36, 37, 42, 44, 46, 61-2, 74, 75, 83, 100, 104, 146
Heathrow airport, London 135
Heenan and Froude, engineering company 17
Heidelberg, Germany 36
Hele Shaw propellor 40
Helensburgh, Scotland 21
Helmsdale, Scotland 86
Hemming, Mr Evans- 53, 66
Henderson, Jock 117

Hendon, airbase and museum, London 42, 81, 117
Henley, aircraft 56
Henschels, aircraft company 36
Hertfordshire 109
Heseltine, Michael 110
Heston Aircraft, company 44, 140
Higgins, Mark 112
Hillard, Mr, headmaster 17
Hillier, Dowty employee 149
Hillman Airways, company 44
Hinkler, Bert 23
Hodson, Mr 58
Hogg, Quintin (Lord Hailsham) 75, 129
Holland 37, 41, 44, 92
Honeybourne, Worcs 14
Hong Kong 106, 107, 113
Houghton. Mrs Johnson 139
Howard, Donald (Lord Strathcona) 89
Howarth, Cedric 67
Hudson, Mr 107
Hudson River, USA, 48
Hue-Williams, Vera 138
Hungary 44, 92
Hunt, R F (Bob) 43, 77, 85, 148
Hunter, aircraft 109
Hurricane, aircraft 54, 56, 109
Hutton, Sir Leonard 85, 151

Iceland 68
Iloman Engineering, company104
Ince, Sir Godfrey 95, 140
India 20, 26, 106
Indiana, USA 98
Indianapolis, USA 98
Ingersoll, Robert 99
Innsworth, Glos 96
Irithlingborough, Northants 20
Irving, Charles and Mrs 26, 109
Isle of Man 41, 46, 58, 93, 104, 106, 115, 143-4, 147-8, 152
Islington, London 19
Ismay, Lady 113
Issigonis, Alec 131
Italy 36, 37

Jackson, Louis 45
Jamaica 128
Jamieson, Mr 66, 67
Japan, Japanese 11, 34, 36, 37, 44, 83, 87, 106–9, 113, 130, 142
Jarvis, Sir Adrian (Jimmy) 140
Jarvis, Jack 137
Javelin, aircraft 79
Jay, Douglas 106

Jeita Grotto, Lebanon115
Jenkins, Walter 40
Jersey 44, 100
Jones, Sir Owen 61

Kansas, USA 49
Kawasaki, company 31, 33, 34, 107, 113
Kearton, Lord 131
Kent, Duke of 52
Kenya 128
Kilner, Fred 81
Kimberley, Canada 89, 90
King, Colonel 97
Kings Norton, Lord 62, 82, 117
Kittyhawk, aircraft 30
Klockner, company 104, 105
Koolhoven, Fritz 19
Kronberg, Germany 36

Lake Constance, Germany etc 36
Lamerie, Paul 139
Lancashire 135
Lancaster, Robert 77, 83-4
Lancaster, Mr 112
Lancaster, aircraft 54, 56, 64, 65, 70
Laxey, Isle of Man 46
Lebanon, Lebanese 99, 113–16
Leeds, W Yorks 53
Leicester 123
Leopard, aircraft 36
Levick, V O 22, 77
Lewis, Sir Edward (Ted) 45
Lidbury, Sir John 117
Lilley, Peter 142
Lincoln, aircraft 56
Lindley, Sir Arnold 104
Lingham, 'Dopey' 140
Lister, Joe 135
Lithuania 44
Liverpool 52, 53, 55, 92
Llewellyn, Colonel 64
Lockheed, aircraft company 49, 119
Longden, David 24
Longden, Henry 92
Lyford Cay, Bahamas 78
Lygon, Dick 135
Lysander, aircraft 38, 47, 56

Madeira 75
Madresfield Court, Worcs 136
Malta 101, 102, 105, 128, 140
Malvern, Worcs 19
Man, Isle of 41, 46, 58, 93, 104, 106, 115,
    143-4, 147-8, 152
Manchester 16, 56, 74

Manhattan, USA 100
Manitoba, Canada 91
Mapplebeck, Mr 115
Marlborough College, Wilts 83, 142
Marshall, Bill 139
Martin, Sir James 109
Martyn, A W 24, 39, 40, 73, 133
Masefield, Sir Peter 69, 82
Mason, Dr Harry 71
Massachusetts, USA 48
Matusch, Mr 42
Maudling, Reginald 86, 88, 95, 104, 109
Mavor, Anderson, company 123
Maxwell, Farnham 139
Mayo, R H 28
McGill University, Canada 89
McKenzie, Arden 21
Meco, company 112
Melville, Mr 11, 30, 32, 33
Merriams Aviation Bureau 28, 29
Messier, aircraft company 30, 62, 64, 96,
    97
Metropolitan Vickers 37, 52, 66, 82
Milford Haven, Wales 106
Milner, Mr 40
Milwaukee, USA 99
Mintoff, Dom 101
Molesworth, Henry 23
Molitor Cup, curling 133
Monckton, Sir Walter 83, 136
Montreal, Canada 69, 77, 141
Morcom, Belliss and, company 19
Morgan, motor car 19
Morrison, W S (Lord Dunrossil) 85, 87
Motherwell, Scotland 123
Moult, Dr 79
Murray, Lee 40

Nabarro, Sir Gerald 136
Napier, aircraft company 21, 22, 62
Nassau, Bahamas 75
Nelson, Sir George 64
Nelson, Peter 139
Netherlands 37, 41, 44, 92
New Hampshire, USA 135, 141
New Mendip Engineering, company 75,
    79
New York 48, 49, 60, 61, 68, 72, 100, 141
New Zealand 19, 98, 128
Newcastle 80
Newey, Clem 15
Newmarket racecourse, Suffolk 138
Nicholson, Frenchie 139
Niedelman, Hilda 60, 72
Niedelman, Sam 48, 58, 60

Nigeria 128
Noorduyn, Bob 19, 69
Norman, Sir Arthur 118
North, J D 62, 105
Northam, Devon 14
Northamptonshire 135
Norway 37, 44
Norway, Nevil Shute 33
Norwich, Norfolk 19
Novello, Ivor 53
Nuffield, Lord 95, 131

Ogden, James 56
Ohio, USA 60
Oklahoma, USA 48, 69
Olding, Jack 139
Olivier, Borg 101
Olley, Captain 93
Olley Company 41
Ombersley, Worcs 21
Orr-Ewing, Ian and Joan 109
Oxford 53, 70, 71, 82, 83, 127, 142

Pakistan 106
Palestine 114
Palmer Tyre Company 111
Panama 91
Paris, France 16, 27, 29, 46, 48, 114, 118,
    146
Parnell Aircraft, company 44
Paul, see Boulton Paul
Pegg, Bill 42
Pena, Paul de la 112
Penang, Malaysia 15
Percival, aircraft company 36
Perrin, Dyson 17
Pershore, Worcs 7, 13–17, 47, 71, 117, 128
Phantom, aircraft 100
Pharaons, Henri 115
Phillips and Powis, aircraft company 36,
    44, 62
Piggott, Lester 139
Pilkington family 75
Poland 37, 44
Popjoy, aircraft company 44
Povey, Mr 40
Powis, Philips and, aircraft company 36,
    44, 62
Prinknash Abbey, Glos 146
Pritchard, J Lawrence 82
Profumo, John 104
Pucky, Sir Walter 46
Pugsley, Sir Alfred 85, 151
Pugsley, Dr 62
Pye Company 46

Rackham, employee 18
Radcliffe Infirmary, Oxford 53
Rambler Cars, company 102
Rapide, aircraft 56
Raymond, Mr 89
Rayrolles, company 80
Remploy, company 83, 85, 95, 136, 151
Renwick, Sir Robert 63, 64, 96
Reykjavik, Iceland 68
Rhine, River 92, 93
Richards, Sir Gordon 139
Rimell, Tom 137, 139
Robbins Committee 129
Robens, Lord 117, 123
Robertson, Lord (Sir Brian) 122
Roe, A V 21-3, 24, 37, 62, 74
Rolls Royce 96, 97, 110, 119, 136, 151
Rolt, L T C 9
Romania 37, 44
Ronaldsway airport, Isle of Man 93
Rootes, Sir William 88
Rose, Billy 48
Rosebery, Lord 137
Rothermere, Lord 39
Rothschild, Lord 126
Rotol, aircraft company 96, 97, 100, 106,
    112, 149, 150
Rowbotham. Mr 64
Rowlandson, Sir Graham 85, 151
Rownson Drew & Clydesdale 20
Roxby-Cox, Dr Harold 62
Rubery Owen Messier, company 62
Rushen, Isle of Man 144
Rutherford, Lord 10
Ryan, aircraft company 49

Salzgitter, company 92
San Diego, USA 49, 50
San Francisco, USA 49
Sandys, Duncan 100, 104
Saro, aircraft company 36
Saskatoon, Canada 89, 91
Saudi Arabia 128
Saunders Roe, company 27, 35, 36, 44
Scotland, Scottish 21, 52, 72, 86, 123, 133,
    139
Seattle, USA 49
Sempill, Lord ('Master of') 28
Serbia 17
Shark, aircraft 56
Shepton Mallet, Som 51
Short Brothers, company 44
Shute, Nevil 33
Siam (Thailand) 36, 37, 44

Sigrist, Fred 75
Sikorsky, helicopter 98
Silverburn Farm, Isle of Man 144
Simmonds, Sir Oliver 11, 63
Simon, Sir John 45
Sjong, Bjarne 37
Skillicorn, Captain Henry 144
Skua, aircraft 56
Skymaster, aircraft 68
Slattery, Sir Matthew 104
Smith Radial Aero Engine 17
Smith, Sir Alan Gordon 28, 139
Smith, Sir Reginald Verdon, 64, 97
Sopwith, Tommy 10
South Africa 44, 92
South America 49, 58, 112, 128
Southampton 21–24, 136
Spain 44, 117
Spartan Cruiser, aircraft 36
Stace, Leonard 41
Stalwart, Alvis, boat 98
Standen, Jim 135
Starfighter, aircraft 49
Staveley Industries, company105
Staverton, Glos 53, 96, 150
Stephens, Gillette 52, 57
Stevenson, George 131
Stevenson, Jordan and Harrison,
    consultants 46, 53
Stewart, Major C J 72
Stirling, aircraft 56
Stockholm, Sweden 40, 102
Stockleigh English, Devon 14
Stonehouse, John 110
Strathcona, Lord 89
Stricklands Roundabouts 13
Sudbury, Canada 89, 91
Sullivan Mine, Canada 90
Sunningend, Cheltenham 25
Surrey 75, 135, 136
Sussex 135
Sweden, Swedes 36, 37, 41, 44, 102
Switzerland, Swiss 44, 54, 56, 83, 106,
    128, 133, 142, 143

Taunus mountains, Germany 36
Tavistock, Devon 110
Taylor, E P 139
Taylor, Enid Stamp 45
Taylor, Jack 99
Tecalemit, company 79
Tenby, Wales 72
Tennessee, USA 99
Tewkesbury, Glos 52, 63, 64, 86, 87, 105,
    113, 117, 128

Thacker, George and Arthur 75
Thailand, see Siam
Thaneshurst, Isle of Man 46
Thatcher brothers 79
Thoma, Professor 96
Thomas, Sir Miles 95
Thompson Produces (Ramo Woolridge)
    78
Thomson, Lord 56, 113, 116
Thornycroft, Lord 88
Tiger Moth, aircraft 27
Timmins, Canada 89, 91
Tinson, Clifford 22, 38
Toba, Japan 107
Tokyo, Japan 35, 107
Tornado, aircraft 56
Toronto, Canada 77, 78, 89
Torquay, Devon 40, 72
Toulmin, Colonel 60
Trevaskis, Henry 111
Trevelyan, Sir Humphrey 126
Tripoli, Lebanon 114
Tupolev TU-144, aircraft 118
Turner, Eric 111
Twickenham, London 133

Ullman, Mr 58–60
United States 4, 55, 60, 68, 69, 82, 100,
    130; see also America

Val d'Esquiere, France 143
Valetta, Malta 102
Vancouver, Canada 89
Versailles, Paris 43
Vickers, Metropolitan 37, 52, 66, 82
Victor, aircraft 79
Victoria Institute, Worcester 18, 20
Villiers, Charles 123
Vimy Ridge Memorial, France 40
Virginia 99
Viscounts, aircraft 82
Vulcan, aircraft 74

Walduck, Stan and Wendy 109
Wallace, Mr 49
Walton, Arthur 81
Walton, Ed and Dodie 141
Wandle Rubber, company 112
Wankel engine 102
Warner, Borg 99
Washington, USA 49, 60, 82
Watkinson, Harold 104
Weatherby's Racing Bank 138
Wegerif, Mr 57
Weir, Lord (J K) 66

Weir Commission 67–8
Welbourn, Donald 126, 127
Welkin, aircraft 56
Wellington, Duke of 131
Wembley, London 135, 139
Wengen, Switzerland 106, 133
West Indies 128
Westinghouse, company 52
Westland Aircraft, company 38, 44, 56
Westminster Abbey, London 54
Westonbirt, Glos 142
Westport, USA 60
Westward Ho!, Devon 14
Whirlwind, aircraft 56
Whittingham, Harry 99
Whittle, aircraft 56, 60
Whitworth, Armstrong 44
Wichita, USA 49
Wigley, Spencer 27
Willesden Junction, London 19
Williams, Dr Lincoln 72
Williamsburg, USA 99
Wilmot Breeden, company 40
Wilson, Gerry 139
Wilson, Dr Henry 72
Wiltshire 64, 75, 79
Wimbledon, London 140

Winchcombe, Glos 41
Winnipeg, Canada 89, 91
Winterbotham and Gurney, solicitors 29
Wolverhampton, Staffs 105
Woolridge, Thompson Ramo 78
Worcester 13, 15–21, 71, 112-13, 118, 129,
    131–37, 146
Worcestershire 9, 13, 15, 21, 113, 117-18,
    133–36, 148
Wragg, Sam 139
Wright, Joe 31, 63, 111, 112
Wright, Wilbur and Orville 30, 82
Wright Field, company 47, 60
Wylie, Major 61
Wynn, Mr 19
Wyre, Worcs 14

Yeovil, Som 38, 56
York 20, 33, 35
York, aircraft 56
Yorkshire 135
Yugoslavia 44

Zambia 128
Zeppelin, Graf 36
Zuckerman, Lord (Sir Solly) 85, 100, 151

Lightning Source UK Ltd.
Milton Keynes UK
UKHW022024170223
417180UK00003B/22

9 781906 978945